UHF RFID Technologies for Identification and Traceability

To Tanguy Laheurte,
student in telecommunication engineering at Telecom Bretagne

FOCUS SERIES

Series Editor Pierre-Noël Favennec

UHF RFID Technologies for Identification and Traceability

Jean-Marc Laheurte
Christian Ripoll
Dominique Paret
Christophe Loussert

WILEY

First published 2014 in Great Britain and the United States by ISTE Ltd and John Wiley & Sons, Inc.

ISTE Ltd
27-37 St George's Road
London SW19 4EU
UK

www.iste.co.uk

John Wiley & Sons, Inc.
111 River Street
Hoboken, NJ 07030
USA

www.wiley.com

Library of Congress Control Number: 2014934510

British Library Cataloguing-in-Publication Data
A CIP record for this book is available from the British Library
ISSN 2051-2481 (Print)
ISSN 2051-249X (Online)
ISBN 978-1-84821-592-4

Printed and bound in Great Britain by CPI Group (UK) Ltd., Croydon, Surrey CR0 4YY

Contents

Introduction

Passive ultra high frequency (UHF) radio frequency identification (RFID) is an electronic tagging technology commercialized between 860 and 960 MHz that allows an object or person to be automatically identified at a distance of up to 10 m without a direct line-of-sight path using a radar-type radio wave exchange (Figure I.1). UHF is the dominant technology for supply chain management applications such as case and pallet tracking and returnable container identification. It is also widely used for real-time inventory, industrial automation, work-in-process tracking, asset management, forklift monitoring, personal identification (ID), vehicle access control, document security and authentication. There are numerous UHF standards, most notably ISO 18000-6 and EPCglobal Gen 2, which are the most widely supported RFID standards nowadays.

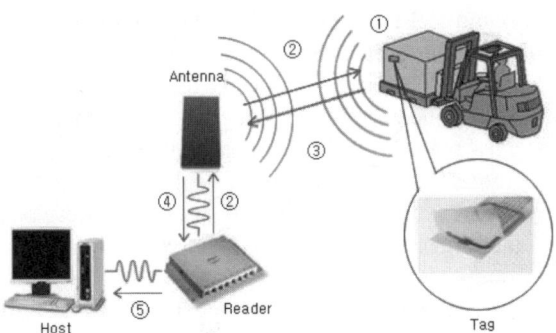

Figure I.1. *Functional principle of a UHF RFID communication. ① Inlay attached to the item: substrate film onto which the antenna and the chip containing item data are combined. ② Radio frequency emitted by the reader installed on a gate, a cashier counter, etc. ③ The tag sends the data in response to the radio frequency (backscattering modulation). ④ The reader antenna transmits the modulated data to the reader. ⑤ The reader decodes the data and sends it to the host computer*

Today, many bar code processes involve bringing the bar-coded object to the reader and orienting the bar code for proper presentation to the reader. As a result, conveyor systems must run considerably slower than their top speeds so that bar code readers can identify passing objects. Unlike a bar code, the RFID tag does not necessarily need to be within the line of sight of the reader and it may be embedded into the tracked object. Moreover, RFID has the ability to identify multiple objects simultaneously. These are the important advantages of RFID over bar code technology, since they eliminate much of the labor currently required and increase the reading speeds.

Apart from the frequency, the main difference between high frequency (HF) and UHF RFID technologies is their read/write distance. The maximum distance for HF technology at 13.56 MHz is approximately 1 m and the maximum distance for UHF technology is approximately 10 m. When long read distances are required, 13.56 MHz technology is not an option: HF antennas do not radiate as their length is very small compared to the wavelength with an almost zero radiation resistance. HF antennas must be seen as near H-field sensors. Most UHF systems communicate by radio waves that provide longer distance. However, UHF systems can also work at close distance and one can find commercialized near-field UHF systems optimized for short-range reading. The same tags that can be read from a 10 m distance can also be read from 10 cm. For instance, UHF technologies can simultaneously satisfy requirements for short-range reading at assembly stations and long-range storage, shipping and receiving processes. UHF systems are then optimized for various processes and reading distances through the placement and configuration of readers.

In early deployments, the UHF systems suffered from performance degradation when used around liquid and metal compared to HF solutions. Advances in antenna design, reader tuning and best practices have overcome this limitation. For example, some UHF tags have been designed specifically for use in close proximity to metal and to take advantage of the conductive properties of the metal to enhance the RF performance. The idea that HF technology is required for use around metal and liquid is nowadays more of a perception than a reality. In supply chain applications, UHF is frequently used to successfully identify cases and entire pallets of consumer goods with high liquid or metal content.

The purpose of this book is not to be exhaustive but to focus on specific aspects of passive UHF RFID technologies. Its first objective is to provide a reference document on the tag antenna design and chip technologies, either from up-to-date academic papers, industrial data or author experience. The second objective is to include perspectives on end users, market and production. Nevertheless, important UHF RFID topics such as the architecture of the readers, the ISO 18000-6 air interface protocol standard and the worldwide regulations are beyond the scope of this book.

The book is organized as follows:

– Chapter 1: Design and performances of UHF RFID integrated circuits (C. Ripoll);

– Chapter 2: Design of UHF RFID antennas and tags (J.M. Laheurte);

– Chapter 3: Design methodology as a function of RCS and DeltaRCS, consequences on the near-field/far-field issues (D. Paret);

– Chapter 4: Markets, applications and end users (C. Loussert).

Chapter 1 provides an up-to-date state of the art in technologies and performances of UHF RFID integrated circuits (ICs). This includes the direct current (DC) voltage supply generation circuit and its regulator, the demodulator part to recover the data, the on-board oscillator to control the digital part and the organization of this digital part and memory. The focus is on the EPC Gen2 protocol adopted in the ISO 18000 Part 6. This added to a full description of the IC functionalities, fabrication issues, matching requirements and measurement tests and benchmarks should help chip designers to identify the main constraints of the technology.

Chapter 2 highlights the design and manufacturing issues of RFID tags. The antenna miniaturization on inexpensive materials is only one of several problems that a designer needs to solve. Tracking fluxes of goods between different companies and across the world must be performed with good read performance despite a close environment characterized by disturbances (associated items, other tags, surrounding objects, etc.). This can only be done if the tag design follows several rules, one of them being the wideband impedance matching to the RFID IC. Another rule is to limit the tag sensitivity to the environment by including the dielectric, conductivity and shape features of the tagged item in the analysis in order to take advantage of it or to fight against it.

As most UHF RFID tags are dipole-based structures, Chapter 2 first describes the fundamental circuit parameters of the dipole antenna (input impedance, radiation resistance, efficiency, Q-factor and impedance). The miniaturization strategies based on fat dipoles, tip loading and meanders are then presented followed by a description of the influence of the dielectric and metallic environments on the tag performance. The fundamental problem of impedance matching between the RFID chip and the antenna is clearly stated with a careful explanation of the single- and double-tuned matching strategies. It is demonstrated that the wide bandwidth characteristic of double-tuned matching circuit is crucial in the presence of dielectrics. Inductively coupled fed tag antennas, as well as the associated commercial loop-based modules, are also extensively detailed. Examples of their use on either filled or empty recipients are given. To conclude this chapter, a state of the art for tags on metal is proposed. Thin and thick structures are examined in succession.

In Chapter 3, special emphasis is put on the design methodology as a function of the radar cross section (RCS) and DeltaRCS with consequences on the near-field/far-field communications.

Chapter 4 is an overview of UHF RFID challenges including the applications, markets, trades and basic technologies, more specifically in the supply chain management and the retail inventory. It is demonstrated that return-on investment (ROI) is key to the RFID buying decision process. RFID technology must generate cost reduction and sales increase to trigger the associated investment. Key topics of future RFID are also detailed: use of tags throughout the whole product life, smart embedded RFID solutions, seamless and ubiquitous infrastructures, and future softwares in massive networks of small intelligent devices.

Design and Performances of UHF Tag Integrated Circuits

Design of UHF RFID tag IC presents unique design challenges to satisfy constraints due mostly to the remote biaising of the batteryless tag. After a brief introduction (section 1.1) and a presentation of the architecture (section 1.2) of a tag IC, section 1.3 will show the principles of converting RF into DC via voltage multipliers successively; first in the ideal then in the real case. The end of the section will deal with the influence of the active element (the diode or the MOSFET, a comparison between the two will highlight the pros and cons of each) and passive parasitics that must be taken into account during the dimensioning of the intermediate and the output capacitors. A simplified model of the antenna and the input of the rectifier will allow us to see the importance of matching and will lead to the computation of the Power Conversion Efficiency (PCE) of the circuit. Sections 1.4 and 1.5 propose a few up-to-date circuits with careful design to reduce the threshold voltage of the active element and improve the PCE. Sections 1.6, 1.7 and 1.8 rather briefly discuss the problem of exchanging information between the reader and the tag and the improvements on the oscillator design to reduce overall consumption. Sections 1.9 and 1.10 list the latest technologies, techniques and trends used in the digital part and lists of performances of the different teams are compared.

1.1. Introduction

The ratification of the global ultra high frequency (UHF) passive radio frequency identification (RFID) standard ISO18000-6 has stimulated the interests of many research laboratories, prompting them to carry out research and development work on the UHF power rectifiers at the microwatt level. In fact, micropower rectifiers are not only limited to RFID but also useful in energy-scavenging modules for remote sensor applications [TEH 09].

The design of an integrated circuit for a UHF RFID tag is not a simple task because it requires numerous constraints to be taken into account.

The primary characteristics of an RFID tag are the cost, the communication range between the tag and the reader/writer, and the transaction time associated with the system performance. To minimize the cost, the tag should be manufactured with the tag integrated circuit (IC) and the associated antenna in a simple process; we will see later that the design rules imply both parts and then each part cannot be designed independently of the other. Despite its simple passive structure, an RFID tag should provide value-added services enabling specific RFID functions, such as data writing, the storing of historical manufacturing or distribution process data, and anticollision reads to speed up the inventory search or security functions to authenticate users [NAK 07].

First, as mentioned above, we must end up with a product for which the cost, so as not to be prohibitive for the retail RFID transponder, should be targeted at being only a few cents. Because the cost of the IC is an important part of the overall cost, it implies the choice of low-cost very large-scale integration (VLSI) technologies, which do not correspond to the best choice for some problematic designs such as the design of the rectifier. Then, a tag IC designer must deal with the challenges of low supply voltage, very low consumption, high input power dynamic range and efficient antenna matching. Because the read range is set by the forward link in a passive backscattering UHF RFID system, it means that the minimum turn-on power for the RF IC chip is of prime importance among the constraints.

A few manufacturers jealously guard their secrets about the design and fabrication process. They sell commercial products with performances as good as the ones displayed by the research laboratories. Some topics such as the optimal choice of the shunt resistor that enables the control of the received power from the far-field to the near-field are not available in the current literature but are actually implanted in certain products. This chapter aims to understand the design principles of the tag integrated circuit, especially the voltage multiplier. Some performances of the power conversion efficiency are also given with respect to different technologies and circuit topologies.

1.2. Integrated circuit architecture

A typical block diagram of a complete passive transponder architecture, including the IC and the matched antenna, is shown in Figure 1.1. Usually, we distinguish between the front-end, which is constituted of the direct current (DC) supply generation, the demodulator and the modulator and the digital part, which includes the control logic, and the electronically erasable and programmable read-only memory (EEPROM) with its charge pump.

The transponder must draw the power required for its functioning from the received electromagnetic field. This power is used mainly by the digital section (often up to 70%) and by the front-end to receive the data sent by the reader and to allow data transmission from the tag to the reader through backscattering modulation.

The regulator circuit stabilizes the output voltage of the multiplier, but it may also keep the input voltage of the multiplier below the breakdown voltage in case of a tag being close to the base station. The voltage reference is sometimes called bandgap reference and output necessary voltages (and currents sometimes) for protection (used by regulator, for example).

1.3. RF to DC conversion: modeling the system

There are two important goals for achieving high power efficiency of the transponder. The efficiency is defined as the ratio between the RF power available at the transponder's antenna and the DC power at the output of the DC block for supplying the transponder. The first goal is the power matching between the antenna and the IC, and the second goal is the RF to DC conversion taking into account the output load constraints, namely a minimum DC voltage to operate the transponder and a minimum load current drawn by the IC (so even if the definition mentions the output power, it is important to note [BAR 09] that each parameter must be independently satisfied). So, one of the big challenges a designer must face is the design of the rectifier with high efficiency while maintaining a minimum DC output voltage and current to supply the transponder.

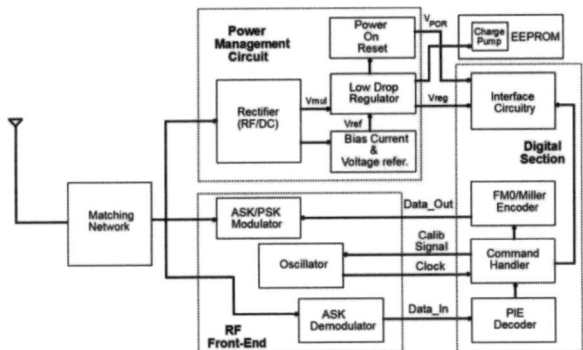

Figure 1.1. *Architecture of a passive RFID transponder*

1.3.1. *Determination of the ideal DC output voltage*

For UHF RFID applications requiring several meters of communication distance, the incoming signal level is only a few hundreds of mV when minimum sensitivity is considered. Therefore, only a multistage rectifier can deal with these requirements and it is used. The topology used by Dickson in 1976 has only been slightly changed by Karthaus and Fischer [KAR 03] in order to make it useful for the alternating current (AC)/DC conversion as shown in Figure 1.3.

The received AC input voltage is converted to a DC output voltage by the voltage multiplier, which is then stabilized and maintained within limits by the voltage regulator [DEV 05].

The elementary cell is built from the clamping circuit C-D1 (see Figure 1.2(a)), which shifts the negative portion of the input signal above zero by storing the equivalent electric charge on the output terminal of C1 by the charging current circulating from ground to IN through the D1 diode. Then, the rectifier circuit detects the peak value of the output signal of the clamp circuit. The electric charge previously stored is now delivered to the output capacity C_{out} by the charging current circulating through D2. When considering ideal elements, we can write the voltage at the output of the clamp circuit [CUR 07]:

$$V_{out}(t) = \hat{V}_{in} + V_{in}(t) \qquad [1.1]$$

where $\hat{V}_{in}$ is the peak value of $V_{in}(t)$, voltage at the input of the multiplier. So, in this idealized model, the maximum possible voltage at the output of the clamp circuit is $2\hat{V}_{in}$. At the output of the rectifier circuit, this value is maintained by the parallel charged capacitor C_{out}.

In the real case, this value is reduced by the voltage drop of the diode.

$$V_{out} = 2\left(\hat{V}_{in} - V_d\right) \qquad [1.2]$$

where V_d is the diode drop voltage.

Besides, this value is further reduced due to imperfections of the circuit elements like the leakage current of the capacitor, the parasitic parallel resistor and the reverse current of the diode.

The half-wave voltage doubler is obtained by cascading the two circuits as illustrated in Figure 1.2(a).

To take advantage of both polarities of the input signal, we must use the full-wave voltage doubler as illustrated in Figure 1.2(b). This implies that the following voltage regulator is able to receive a differential input.

To reach the necessary output voltage (which depends on the complementary metal oxide semiconductor (CMOS) technology used but is actually approximately

1.2 V) when the tag is in the far-end, it is mandatory to use an N-stage multiplier, which consists of a cascade of N elementary cells.

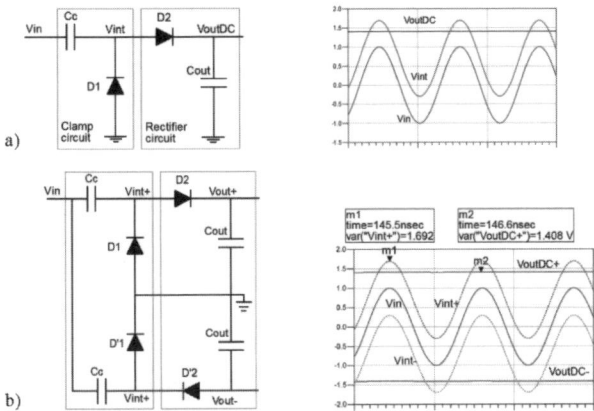

Figure 1.2. *Elementary cell of an N-stage multiplier: a) half-wave voltage doubler and b) full-wave voltage doubler*

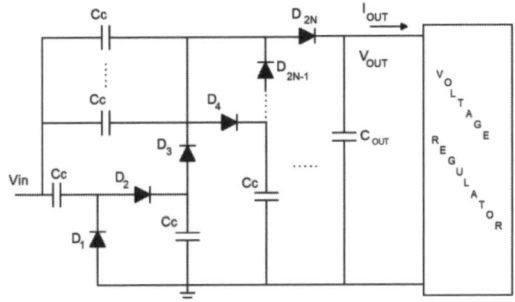

Figure 1.3. *N-stage half-wave voltage multiplier and voltage regulator*

Then the voltage generated between the input and the output for an N-stage half-wave multiplier is:

$$V_{out} = 2N \cdot \left(\hat{V}_{in} - V_d \right) \qquad [1.3]$$

In the DC analysis, capacitors act as open circuits, so we now have 2N identical diodes in series; so the voltage drop across each diode may be written with respect to time as:

$$V_d(t) = \pm V_{in} \cos(\omega_0 t) - \frac{V_{out}}{2N} \qquad [1.4]$$

1.3.2. *Determination of the "real" DC voltage*

Actually, equation [1.3] is a crude approximation because the threshold voltage is considered as constant. In fact, it depends on the direct current I_d and saturation current I_s through an exponential law between current and voltage for the Schottky diode (or a square law in the case of a diode-connected MOS); so we have a forward diode drop that is logarithmically dependent on the diode current, which is actually the load current:

$$V_d \approx \eta \cdot V_{thermal} \cdot Ln\left(\frac{I_d}{I_s}\right) \qquad [1.5]$$

where η is the diode non-ideality factor. So, it should be rewritten as equation [1.6] to take into account this dependency but also the choice of the technology through the saturation current:

$$V_{out} = 2N \cdot \left(\hat{V}_{in} - V_d(I_d, I_S)\right) \qquad [1.6]$$

In UHF RFID applications, the amplitude of the input voltage is rather weak and the diode operates in a region where the voltage drop depends strongly on the current as shown in Figure 1.4.

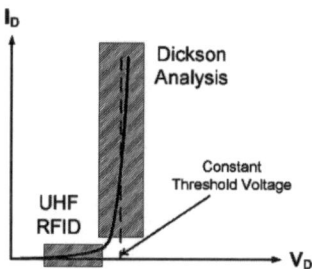

Figure 1.4. *Diode operational regions for Dickson's original analysis and for UHF RFID application (after [BAR 09])*

For example, if I_S is 200 nA and the current through the diode varies from 2 to 4 µA, then the voltage drop will vary from 60 to 75 mV.

The diode current has a pulsed shape due to the nonlinear relationship (equation [1.5]) as shown by some simulations [BAR 09] in Figure 1.5 for a coupling capacitor of 1.2 pF, an output capacitor of 12 pF and a diode saturation current of 120 nA.

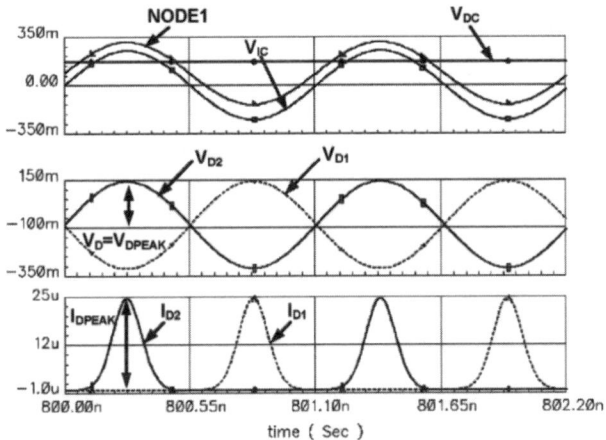

Figure 1.5. *Voltages and currents for the Schottky diode doubler in steady-state conditions (after [BAR 09])*

As can be seen, the threshold voltage cannot be neglected because it represents a drop approximately 100 mV (depending on the DC), which is of the same order of the amplitude of the input voltage. It is important to find the threshold voltage precisely because a small variation in it gets multiplied by the number of stages, which can significantly change the DC-generated output.

The output of the rectifier could be considered as a current source, whose value is determined by the current consumption of the IC. This means that, because of the charge conservation, the average value of the instantaneous current over one conduction cycle is equal to the load current drawn from the output, that is I_{out} equals 3.4 µA for this example [BAR 09]. So, if the output (or load) current increases, the peak current also increases to allow the increase of the average current as shown in equation [1.7]:

$$\frac{1}{T}\int_0^T I_d(t)dt = I_{out} \tag{1.7}$$

Similarly, if the input voltage increases, the peak diode current will increase as well, and thus the pulse shape will change.

So, from the viewpoint of the DC output generated, the voltage drop in the diode depends on the input voltage and the output current drawn. In fact, because of the way the rectifier works, the threshold voltage is determined only at the peak of the current $\hat{I}_d$.

So, the DC output voltage should finally be expressed by:

$$V_{out} = 2N \cdot \left(\hat{V}_{in} - V_d (\hat{I}_d (\hat{V}_{in}, I_{out}), I_S) \right) \qquad [1.8]$$

Based on the development of the relationship between current and voltage with modified Bessel function series of the exponential of a cosine function, some authors like De Vita and Iannaccone [DEV 05] establish a general N-stage rectifier input–output relationship that can be solved as:

$$\left(1 + \frac{I_{out}}{I_S} \right) \exp \left(\frac{V_{out}}{2\eta N V_{thermal}} \right) = B_0 \left(\frac{\hat{V}_{in}}{\eta V_{thermal}} \right) \qquad [1.9]$$

Here, we clearly see that a designer can decide to choose V_{out} as the objective output parameter, leaving I_{out} as a dependant variable [TEH 09].

1.3.3. *Effects of parasitics and capacitances on the output voltage*

So far, we have only taken into account the threshold voltage as the main parameter to determine the DC output voltage but, in fact, other parameters can have a secondary influence, namely parameters of the diode model, the coupling and hold capacitors.

First, we need to know what the non-idealities are and where the parasitics for each element of the rectifier are.

1.3.3.1. *Active elements parasitics*

For the charge transfer devices acting as a switch, the designer has the choice between the Schottky diode, first introduced by Karthaus and Fischer in 2003 [KAR 03], and the diode-connected metal oxide semiconductor field effect transistor (MOSFET). For the latter, many alternate solutions have appeared such as the use of native transistor, special biasing circuitry, threshold programming by analog memory or the dynamic threshold MOSFET [TEH 09].

In designing a Schottky diode for a UHF RFID system, value, associated nonlinearity and the RC cutoff frequency of the diode are of primary concern. Moreover, the parasitics of the diode depend on the geometry and physical process. The way to implement diodes on a standard CMOS process must be done for an easy integration compatible with the usual masks used. Figure 1.6 illustrates a typical planar diode [JAM 06b] and its associated electrical model in Figure 1.7. It consists of a Schottky contact (Ti) to an n-well active zone and an ohmic contact to a heavily doped n$^+$ layer. This structure allows a compact layout so as to minimize the number of extrinsic parameters and especially the capacitors.

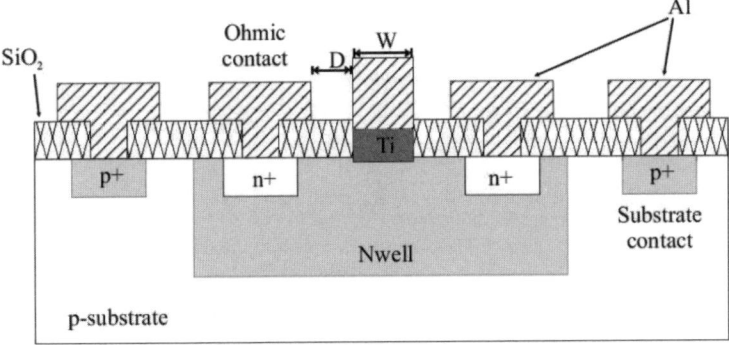

Figure 1.6. *Planar Schottky Barrier diode cross-section (after [JAM 06b])*

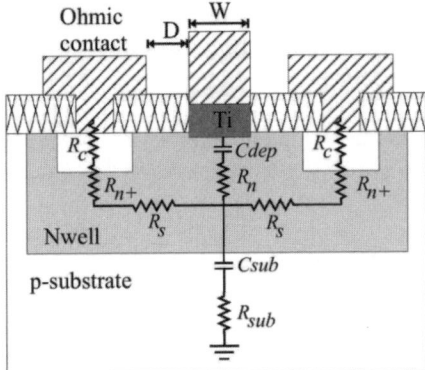

Figure 1.7. *Electrical model of the planar Schottky barrier diode cross-section(after [JAM 06b])*

When considering the choice of the Schottky diode, it is important to have a large saturation current I_s that will result in low forward voltage drop (see equation [1.5]), a small junction capacitance C_D as well as small parasitic capacitance to substrate C_{Tub}. The problem is that a large Schottky diode has a large saturation current but a large capacitance also, which actually dominates the power losses, and an optimum choice of the diode has to be found.

At intermediate node NODE1 in Figure 1.8, we clearly see that this substrate capacitance behaves topologically as a parallel capacitance $C_{PARA,}$ which is the sum of the junction capacitances C_{D1} and C_{D2} plus the "tub" capacitance of the series diode D_2 labeled C_{TUBD2} plus the interconnect wires capacitance. C_{PARA} is typically an order of magnitude smaller than C_c.

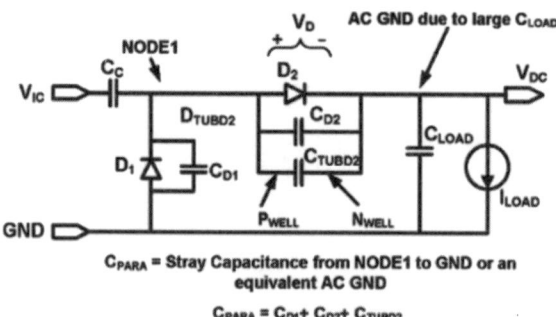

Figure 1.8. *Schottky diode cross-section and doubler with its parasitics (after [BAR 09])*

The series resistor of the diode and of the coupling capacitor can be minimized by using a multi-finger structure. Very often, the series resistor Rs is neglected in UHF RFID because the direct diode current remains low (μA), thereby inducing a low resistive drop voltage less than 1 mV.

One of the main limitations of using Schottky barrier diode (SBD) solution is the need to specially modify the CMOS process. Many laboratories have been searching for how to use the MOSFET optimally. As the size of the transistor continues to scale down, the effective threshold voltage approaches the SBD turn-on voltage [TEH 09] approximately 150–200 mV.

The difference with SBD is that the MOS actually operates in different regions in one cycle. When it is on, most of the time it is in the superthreshold region, where the drain current varies as the square of the gate-source voltage. When it is off, unfortunately, it conducts a reverse leakage current (inversion of drain and source). When the gate is connected to the source, this current is the subthreshold current and cannot be neglected because [YI 07]:

– it increases exponentially with the decrease in threshold voltage;

– it can be the same order of the load current (μA);

– the power consumed is not negligible because it remains in this state about half of the cycle.

This is illustrated in Figure 1.9 where, as not shown in Figure 1.5, this reverse current induces a power dissipation.

1.3.3.2. *Passive elements parasitics*

In the same way, the leakage capacitance to substrate of the coupling capacitor offers a parasitic path for the charges and, consequently, it should be minimized.

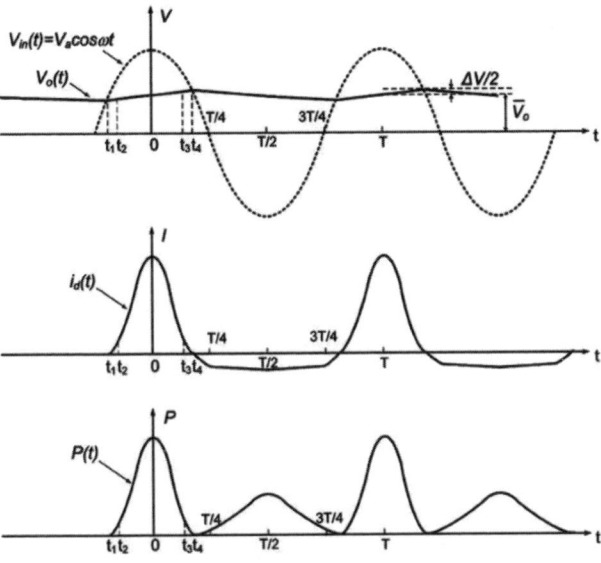

Figure 1.9. *Waveforms of input and output voltage, current and power dissipation (after [YI 07])*

Indeed, the impact of these parasitic parallel capacitors to ground is very important. We will see later its impact in the matching issue; for now, they act as voltage dividers, as shown in Figure 1.8, and thus in the conversion equation, we must take into account not the input voltage but the modified voltage at node 1 (intermediate node).

If we consider the capacitance model as shown in Figure 1.10 [MAN 07], let us suppose that the input signal is connected to the top plate of the coupling capacitor C_c. Ahead in the model, the bonding pad and package capacitances are independent of the size of the rectifier circuit and can be grouped into C_{pak}. The input capacitance of the rectifier is C_{PARA}, which has already been introduced and takes into account the parasitic capacitance to ground at the input of each stage due to the diodes or transistors. The capacitance between the bottom plate of C_c and ground is C_{CP} (typically several tens of femtofarads in design foundry manuals).

This model leads to equation [1.10], which clearly shows the effect of voltage divider depending on the values of C_{PARA}, C_C and C_{CP}, respectively.

$$\alpha = \frac{V_{NODE1}}{Vin} = \frac{C_c}{C_{PARA} + C_c + C_{CP}} \qquad [1.10]$$

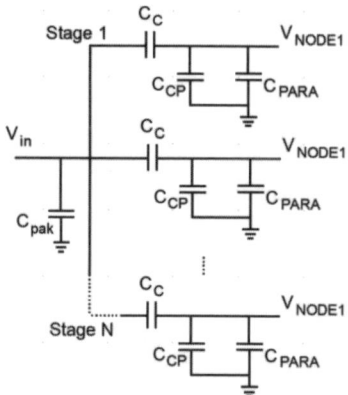

Figure 1.10. *Circuit model of the parasitics of the capacitances before the ideal rectifier (after [MAN 07])*

1.3.3.3. *Effect of parasitics on output voltage*

Equation [1.10] shows that it is very important to reduce the C_{PARA} capacitance. Moreover, the total input capacitance clarifies the fact that all the stages are in parallel. The consequence of this parallelism is that the RF voltage applies to a much lower impedance, so we can anticipate an optimum number of stages for the multiplier with respect to the criteria of the amplitude of the input voltage.

Figure 1.11 illustrates the overwhelming effect of the junction capacitance of the diode in the term C_{PARA} that clearly demonstrates the importance of minimizing the size of the diode, irrespective of the technology used, Schottky diode or diode-connected MOS transistor.

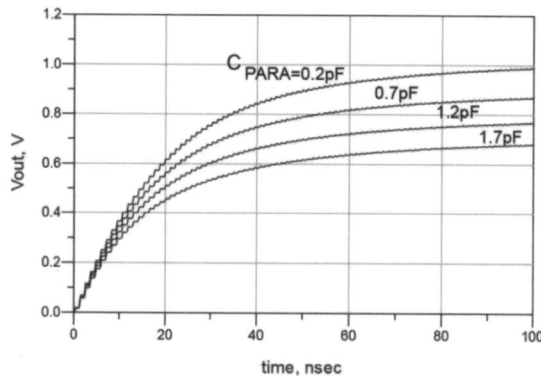

Figure 1.11. *Effect of the diode junction capacitance on the output voltage with C_c=5 pF*

1.3.3.4. *Dimensioning of the intermediate capacitors*

These capacitors act as charge transfer devices. They are charged during the negative cycle of the input signal and then transfer their charge to the output during the positive part of the input signal. Therefore, a small value will decrease the time to transfer the charge to the next stage. On the other hand, a large value would increase the RC time constant; the multiplier acting as a low pass filter for the input signal that can slow down the exchange of data between the reader and the tag. As a compromise and to avoid these intermediate capacitances to act as voltage dividers, it is currently admitted that these intermediate capacitors must be chosen so that they satisfy $C_c \geq 10 \cdot C_D$, so typically a value approximately 2–3 pF. From equation [1.10], in this case, the loss in voltage is not more than C_{CP}/C_C.

As we have seen that the DC output power depends on the input voltage applied to the multiplier, we must consider the behavior of the circuit in the AC mode. In an AC analysis, the intermediate capacitors (all except the output capacitor) should behave as short circuits; therefore, all diodes appear in parallel to the input.

So, for these capacitors, the capacitance to substrate must be reduced. The best way to reduce it is to build them by choosing a multi-finger top metal layer configuration or making use of the capacitance between the top two metal layers [JAM 06a].

1.3.3.5. *Dimensioning of the output capacitor*

To ensure that the output DC voltage is constant, the output or hold capacitor C must be dimensioned so that its time constant is much larger than the period of the RF input signal [DEV 05] and thus guarantees normal operation even when the RF power from the antenna is not available for a time T_{low}. If we consider that most of the load on the rectifier is due to the digital part with an on-board clock frequency T_{ck}, then the minimum allowable value of C_{out} is given by [MAN 07]:

$$C_{out,\min} = C_{sw} \frac{T_{low}}{T_{ck} \cdot Ln\left(\dfrac{1}{\alpha}\right)} \qquad [1.11]$$

where C_{sw} is the total capacitance being switched between V_{dd} and ground by the digital load and α is the percentage of the V_{dd} drop.

This formula leads to current values approximately 50–150 pF.

1.3.3.6. *Effect of parasitics on power loss*

All the dynamic losses due to coupling to the substrate (see Figure 1.7) can be determined using equation [1.12] [KAR 03]:

$$P_{sub} = \frac{1}{2}V_{in-sub}^2 \cdot \frac{R_{sub}}{R_{sub}^2 + (\omega C_{sub})^{-2}} \approx \frac{1}{2} \cdot V_{in-sub}^2 \cdot R_{sub} \cdot (\omega C_{sub})^2 \qquad [1.12]$$

where

V_{in-sub} : device RF peak voltage with respect to substrate;

C_{sub} : total capacitance to substrate;

R_{sub} : total series resistance to substrate.

The assumption made is valid for common low-resistivity substrate where $R_{sub}C_{sub}\omega \ll 1$.

Before proceeding to the final results, we must treat the second constraint, which is the matching.

1.3.4. *Matching considerations*

1.3.4.1. *Voltage and power at the input of the IC*

The effectiveness of the power transfer between the reader and the tag can be improved by maximizing the power transfer between the tag antenna and the rectifier input structure.

The complete system can be modeled by considering an antenna connected to the IC through a matching circuit, as shown in Figure 1.12.

Before calculating the input voltage and power at the IC terminal and deriving the input impedance of the tag, we must consider the transformation aspects brought by the matching network.

According to Barnett [BAR 09], there is no need to use a classical L-matching cell with an inductive series element because the required ratio R_{in}/R_a must be high to bring the Q voltage boosting at the transponder input. But, at the same time, the radiation resistance must be high to increase V_a as will be demonstrated; so, series matching is not the preferred way. On the contrary, parallel inductor matching brings simplicity (same printed technology) and an improved performance to electrostatic discharge (ESD) performance. So, a simple parallel inductor easily printed at the same time as the antenna compensates for the IC input capacitance.

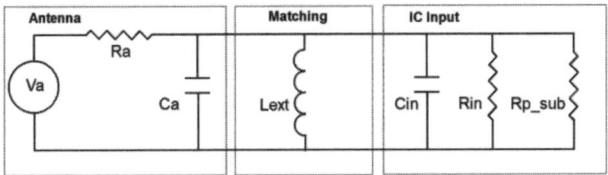

Figure 1.12. *Simplified model of the antenna with matching and rectifier input*

If we define P_{av} as the power available at the antenna terminal when the system antenna-rectifier is matched, we can write:

$$P_{av} = P_{ERP} \cdot G_R \cdot \left(\frac{\lambda}{4\pi d} \right)^2 \qquad [1.13]$$

where P_{ERP} is the effective radiated power and d is the separating distance.

If V_a is the antenna open-circuit voltage and R_a is the antenna radiation, then V_a is related to the power available by:

$$\hat{V}_a = \sqrt{8 P_{av.} R_a} \qquad [1.14]$$

So, we clearly see that the peak voltage V_a is proportional to the square root of the radiation resistance and thus, on this criterion, we must choose an antenna with a high radiation resistance (300 Ω for a folded dipole, for example).

If we assume that the input capacitance of the IC is compensated by the parallel inductor and that the resistance R_{in} represents the equivalent resistance of the input resistance of the multiplier in parallel with the resistance that represents the DC consumption of the entire IC.

In this case, the voltage at the input of the IC can be written as:

$$\hat{V}_{in} = \hat{V}_a \frac{R_{in}}{R_{in} + R_a} = \sqrt{8 P_{av.} R_a} \cdot \frac{R_{in}}{R_{in} + R_a} \qquad [1.15]$$

so, maximum power P_{in} is absorbed by the rectifier when the antenna is matched to the rectifier, and we have $P_{in} = P_{av}$.

To have an idea of the voltage and power obtained in the case of the European regulations, let us make simple calculations of the resulting input power and voltage at the IC terminal in the two cases, $R_a = 300\,\Omega$ and $R_a = 800\,\Omega$ with the following parameters. Note that we do not vary the input resistance of the IC, which is often approximately 1 kΩ but decreases when the drawn DC current increases:

$$F = 868\,\text{MHz} \qquad P_{ERP} = 2\,\text{W} \qquad G_R = 1.64 \qquad R_{in} = 800\,\Omega$$

	Distance	4m	8m		Distance	4m	8m
	Pin (dBm)	-5	-11		Pin (dBm)	-8,5	-14,5
Ra=300Ω	Vin (mV)	440	220	Ra=800Ω	Vin (mV)	480	240

Table 1.1. *Input power and voltage at the transponder IC*

Table 1.1 clearly shows that to satisfy the two criteria of maximum power transferred and, at the same time, maximum input voltage, we must power match the antenna and the transponder IC and must have the highest antenna radiation resistance. This means that the designer needs to reduce the power consumption (choice of technology) and the input capacitance (choice of technology, number of stages and careful layout).

In these calculations, we have supposed that the backscatter modulator is removed and that only the transponder input impedance is taken into account.

As seen before, the power at the transponder antenna varies as the square of the distance between the reader and the transponder. The matching should be considered in the condition of minimum power available at the antenna to ensure correct operation of the tag, that is at largest distance (maximum operating range). Concerning the problem of deciding in which state of the input IC impedance (that is depending on the current consumption Iout) we should do this matching, the answer has been brought by Curty [CUR 07]. He showed that R_{in} should be considered at its minimum (output current at its maximum, so the transponder is fully functioning, both the analog and digital parts), so V_{in} is at its highest level when mismatch between R_a and R_{in} occurs.

1.3.4.2. *Input equivalent impedance*

In addition to the output voltage equation, an equation of the input impedance is important for optimizing the rectifier. In the aim of matching, we need to understand the physical meaning of Z_{in}. First, we can consider two contributions to this impedance as illustrated in Figure 1.12. The impedance of the transponder includes a part that does not depend on the input voltage V_{in} and a part that is nonlinear and thus depends strongly on the input voltage and on the load current.

Let us first investigate the power consumption. DeVita and Iannacone considered the input power required to obtain a given output voltage and power by summing up the average power dissipated in each diode and substrate and the power required by the load P_{out}. Following the analysis of equation [1.9], they showed that the input power can be written as a function of a first-order modified Bessel function [DEV 05]:

$$P_{in} = 2N \times I_S \ \hat{V}_{in} \ B_1\left(\frac{\hat{V}_{in}}{V_T}\right)\exp\left(-\frac{V_{out}}{2NV_T}\right)+P_L+P_{sub} \qquad [1.16]$$

Note that the input voltage considered should be in fact V_{node1} as mentioned in equation [1.10]. So, the input resistance should be seen as an equivalent resistance linked to the power consumption:

$$R_{in} = \frac{\hat{V}_{in}^2}{2P_{in}} \qquad [1.17]$$

We have seen that the relationship between the diode current and the diode voltage was nonlinear. So, here, we make an approximation, when we consider only the first term in the equivalent polynomial equation relating this absorbed current to the input voltage or input power. Even if this is far from reality because of the pulsed current (see Figure 1.5), it seems that considering resistance as the resistance that consumes a mean power P_{in} is enough; this approximation gives a good agreement with the measured results.

For the extraction of the equivalent input reactance, we need to know that this capacitance is a function of the voltage applied to it; so, an averaged value should be taken for a single diode as:

$$\bar{C}_d = \frac{1}{2\hat{V}_{in}} \int_{V_{d\,min}}^{V_{d\,max}} C_d\left(V_d\right)dV_d \qquad [1.18]$$

where $V_{d\,min}$ and $V_{d\,min}$ are the minimum and maximum voltage drop, respectively, and the total input capacitance will be, remembering the parallelism of all diodes:

$$C_{eq} = 2N \times \bar{C}_d \qquad [1.19]$$

Then, when considering the parasitics, we can write the total capacitance at the input of the voltage multiplier:

$$C_{in} = NC_c \frac{C_{PARA} + C_{CP}}{C_{PARA} + C_C + C_{CP}}$$ [1.20]

where $C_{PARA} = C_{D1} + C_{D2} + C_{TUBD2}$

Actually, the resistive part must be modified because we must add the parasitics that come from the loss power in the substrate. It can be physically modeled by a series resistor R_{sub} (see Figure 1.7) representing a parasitic tub resistance. This circuit is converted to an equivalent parallel model [NAK 07]. So, in the end, we have an imaginary reactance due to the parasitic and diode capacitances and a real part that depends on two contributions, one of which is the current consumption and the other is due to resistive and reactive parasitics of the substrate as illustrated in Figure 1.12.

Ideally, Rp_sub should be at least 10 times the value of the antenna resistance, which is approximately 1 kΩ or less than that (but should be maximized in any case to obtain the maximum V_{in}). Actually, its value is dominated by the input capacitance. Furthermore, even if the input capacitance is more or less resonated by the fixed external inductor, it must be controlled because too high a capacitance means an increase of the system Q, which in turn may bring difficulty in covering the 100 MHz bandwidth around 900 MHz. For this reason, the Q is usually limited to 6–8.

EXAMPLE.–

If the operation frequency is 868 MHz and C_{in} is only 1 pF, then the parallel parasitic input resistance can be raised up to only 3 kΩ. At the same time, when the chip consumes 100 μA (analog and digital parts are fully functioning), the input resistance due to consumption is 10 kΩ under a DC voltage of 1 V. So, we can conclude that the required design constraint for the input impedance is to reduce the input parasitic capacitance (small diode or transistor and small substrate capacitance in the well-chosen topology).

For a commercial circuit, it is possible to find these figures: total series input resistance of 6.7 Ω, C_{in} = 0.88 pF, which gives a total parallel input resistance of 5.8 kΩ.

In conclusion, before showing the results obtained, we can say that a large diode (or diode-connected MOS) will show a low threshold voltage V_d and is beneficial for increasing output DC voltage given through $V_{out} = 2N \cdot (\hat{V}_{in} - V_d)$. Also at the same time, a large diode results in large parasitic capacitance and the coupling capacitance should be increased to reduce the loss (see equation [1.10]). However, increasing C_c increases its own parasitic capacitance to ground and adds to the input capacitance of the rectifier. In fact, the diode size and the coupling capacitor should be scaled together to increase the input voltage. Further increase in voltage can be obtained by adding stages with the associated benefit of a reduction of the input resistance that eases the matching with the antenna.

1.3.5. *Results obtained*

Barnett has made an interesting approach that takes into account the previously mentioned capacitive voltage divider and highlights the importance of relating the peak diode current to the load current through a variable χ, $\chi = \hat{I}_d / I_{out}$:

$$V_{out} \cong N \cdot \left(\hat{V}_{in} \cdot \left(\frac{C_c}{C_c + C_{para}} \right) - \eta \cdot V_T \cdot \ln \left(\frac{\chi \cdot I_{out}}{I_S} \right) \right) \qquad [1.21]$$

The assumption made is that the output or hold or load capacitance is much greater than the coupling capacitance, so it does not appear in the equation. This model seems to give a good matching (<5% error at maximum output current I_{out}) for increasing small currents with the nonlinear spice simulations, as illustrated in Figure 1.13.

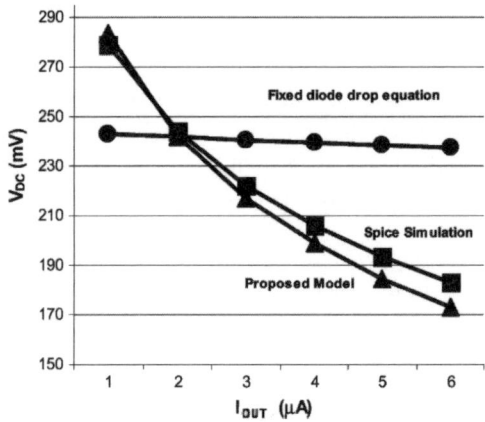

Figure 1.13. *Comparison between the model of equation [1.9], the nonlinear spice simulations and the model with a fixed diode drop (after [BAR 09])*

It clearly shows the strong dependence of the output voltage on the load current, more than 50% of variation for V_{DC} for a 1–6 variation of the small load current. For strong currents (more than 20 μA), the slope decreases. So, when designing a voltage multiplier, we must take into account this fact and consider the necessary turn-on voltage as the smallest (IC is fully operating, load current approximately 80 μA).

By choosing an appropriate number of stages, any voltage can be reached. However, this can be considered valid only for a small current draw. As soon as the current increases, there is a simultaneous AC current through the capacitors, resulting in a voltage drop and a lower input voltage for the subsequent stages. In reality, there are hardly any circuits with more than 15 stages [JAM 06b].

The power conversion efficiency (PCE) is defined by the output power divided by the input power. The input power is, as seen in equation [1.17], the sum of the output power plus the losses in the diode and the loss in the substrate.

$$PCE = \frac{P_{OUT}}{P_{IN}} = \frac{P_{OUT}}{P_{OUT} + P_{LOSS} + P_{SUB}} = \frac{P_{OUT}}{P_{OUT} + N \cdot P_{DIODE} + P_{SUB}} \qquad [1.22]$$

where P_{DIODE} is the power loss of each diode: $P_{DIODE} = P_{FORWARD} + P_{REVERSE}$

Diode losses originate from the resistive loss when current flows through the diode in the on-part of the cycle (determined by the threshold voltage) and from the reverse leakage current when the diode is in the off-part of the cycle. So, a small threshold voltage associated with a small reverse current and a small parasitic input capacitance allows us to reach high PCE.

In Figure 1.14 [KOT 09], we show the PCE as a function of input power for a large input dynamic with respect to different parameters. This study shows some comparisons with other topologies introduced in the next section like the static cancellation technique of the threshold (self-threshold-voltage cancellation (SVC) and external threshold-voltage cancellation (EVC)) and active cancellation technique (internal threshold-voltage cancellation (IVC) and the study).

It is claimed that the threshold voltage V_{th} can be minimized in the forward bias condition and increased in the reverse bias condition, thereby reducing the leakage reverse current.

So, in conclusion, the RF to DC equation should include the effect of the nonlinear forward voltage drop in diodes and the impedance matching conditions between antenna and rectifier input showing the important parameters for the designer, namely:

– the available power;

– antenna radiation resistance;

– number of diodes (stages);

– DC voltage and load current;

– parasitic resistive loss components;

– diode and capacitor sizes;

– operation frequency.

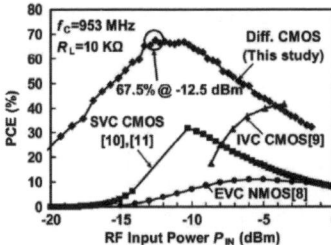 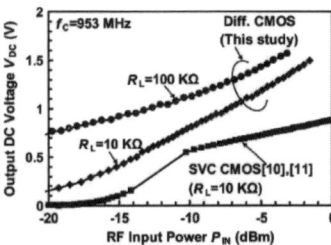

Figure 1.14. *PCE and DC output voltage as a function of RF input power (after [KOT 09])*

1.4. RF to DC conversion: proposed circuits and performances

1.4.1. *Threshold-voltage cancellation circuit*

The traditional rectifier structure based on the Dickson multiplier technology suffers from low power conversion efficiency due to the forward voltage drop in the diode or diode-connect transistor. Despite the fact that a lot of circuits implement the classical structure seen before, either with Schottky diodes or with diode-connected MOSFETs, some other solutions have been proposed. The case of using a zero-threshold transistor has not been the solution due to the significant amount of the reverse leakage current that seriously degrades the efficiency. Although process enhancement helps solve the problem but is usually insufficient, a variety of circuit design methods have attempted to achieve better performance keeping the standard process, by introducing a gate bias to reduce the effect of V_{th}, which is almost equal to the turn-on voltage.

Umeda [UME 06] introduced first its EVC followed by Nakamoto [NAK 07] introducing the IVC.

The rectifier proposed by Nakamoto *et al.* (Fujitsu) is a single-stage mirror-stacked architecture as shown in Figure 1.3; so, it differs largely from the traditional architecture where the threshold voltage is not electrically compensated. The elementary cell is shown in Figure 1.15.

In this circuit, the cancellation of the threshold voltage is achieved by inserting an internal V_{th} cancellation circuit (IVC) between the series PMOS M_{p1} and the output V_{DD}. Capacitor C_{bp} holds the threshold voltage of the PMOS diode M_{p1} by replication of it with M_{pb}. This is done in the same way for the NMOS diode M_{n2}. This circuit is reported to accurately track the process and temperature variations by matching of the transistors. R_b should be chosen as being large enough to avoid the IVC branch currents.

Figure 1.16 illustrates the entire full-wave rectifier circuit. It exploits both polarities as mentioned previously to optimize power efficiency and the mirror structure contributes to eliminating the effect of the parasitic capacitances at the IN-node, which operate as the AC ground.

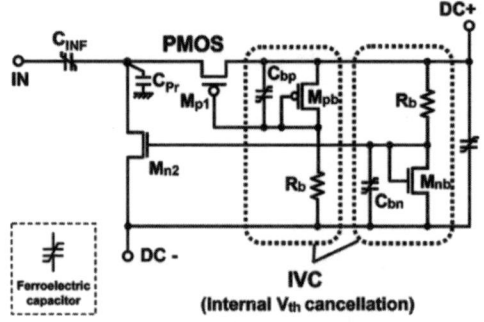

Figure 1.15. *CMOS half-wave rectifier (after [NAK 07])*

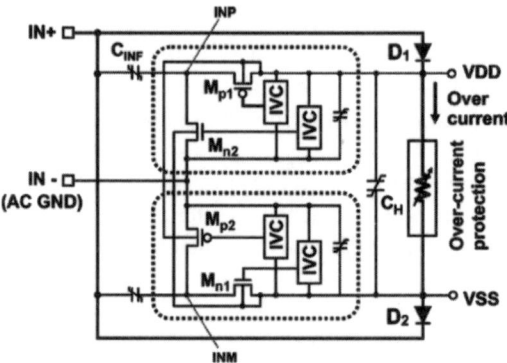

Figure 1.16. *Entire full-wave rectifier with mirror stacked architecture (after [NAK 07])*

The authors report a 36.6% efficiency at 953 MHz for this rectifier for an input power of −6 dBm.

1.4.2. *Cross-coupled differential drive with automatic bridge structure cancellation circuit*

Unfortunately with the previous circuit, when the effective threshold becomes too small due to an excessive DC bias voltage, the MOS transistor can be on for too long a time and an increased reverse leakage current appears, removing the stored charges on the output capacitor. To solve the problem, Mandal [MAN 07] and Kotani [KOT 09] brought an improvement by introducing a dynamic cancellation technique, as illustrated in Figure 1.17. They proved that it is not possible to achieve a small on-resistance and a small reverse-leakage current at the same time with static cancellation circuits.

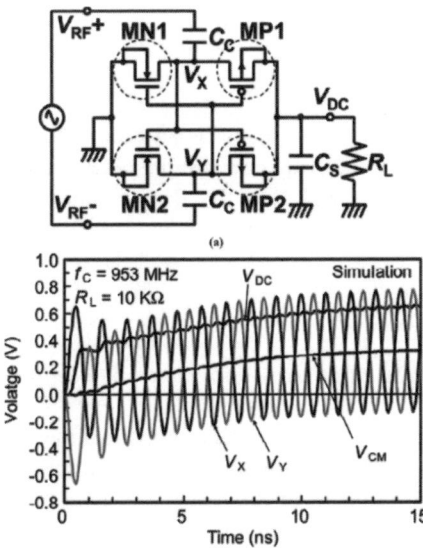

Figure 1.17. *Cross-coupled differential drive with bridge structure (after [KOT 09])*

The simulation result shows a common-mode voltage (DC components of V_x and V_y) which is about half the DC output voltage is generated by the rectification operation and thus is similar to the previous schemes of static compensation. In addition to this, in this differential structure, the gate potentials depend on the differential input signal. By changing the gate polarity of MN1 (and the others of course) when V_x is either positive or negative, the resulting Ron, as well as the reverse leakage current, is reduced.

Figure 1.18 shows the measured results for the static and dynamic types of cancellation techniques, as well as the simple diode-connected transistor.

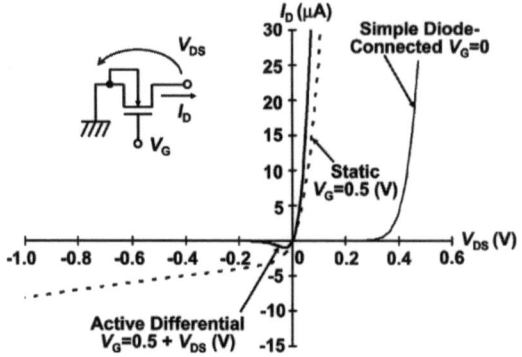

Figure 1.18. *Current-voltage characteristics of diode-connected*
n-channel MOS transistor (after [KOT 09])

DC bias for both types is set to 0.5 V as an example. As mentioned before, the on-resistance is reduced by the static case but the reverse current is increased. In the dynamic case, both are minimized. Note that the generated voltage is:

$$V_{outDC} = 2N \cdot \left(\hat{V}_{in} - V_d \right)$$

$\qquad\qquad$ [1.23]

1.4.3. *Cross-coupled differential drive with controlled tuning voltages*

In the usual methodology, the starting point is the dimensioning of the transistors to satisfy the load current requirement. Some authors like Wong and Chen [WON 11] separate the problem of efficiency and load current by setting a set of circuit parameters (V_{out}, V_{out}/V_{inpeak}) and the transistor width ratio between the PFET and NFET pair). First, they find the set for maximum power conversion efficiency and, subsequently, the transistors are scaled independently in order to meet the load requirement, while keeping the set of parameters constant.

In their rectifier shown in Figure 1.19, they have chosen a cross-coupled differential drive with a bridge structure as we have seen before. All the bulk of NMOS transistors are connected to ground, whereas the bulk of PMOS transistors are differently biased according to their stage number. So, it is a different scheme from what we have just seen with the dynamic procedure where the PMOS gates were connected to the cell's output. It has the advantages of a fine-tuning of each of these voltages according to a digital dynamic control. Note that the capacitances C_{int} could be omitted because the forward current of Q_3 injected into the interstage is simultaneously sunk by Q_8. This holds in case of perfect match between the transistors. Besides, even a small capacitance helps reducing the voltage ripple.

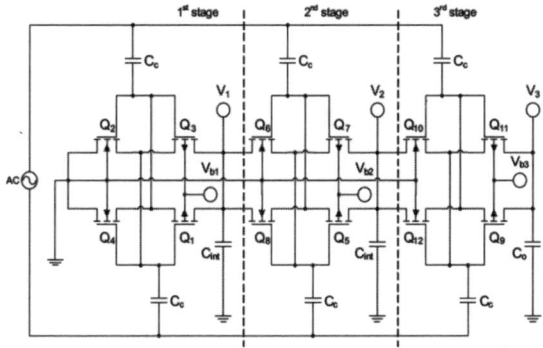

Figure 1.19. *Three-stage differential-drive CMOS rectifier after Wong and Chen*
(after [WON 11])

1.4.4. *Results*

To illustrate the influence of the main design parameters and give an idea of the results one can obtain, we illustrate with the results given by Kotani [KOT 09] (see Figure 1.14).

An efficiency (assuming perfect matching) above 50% is obtained between −16 and −7 dBm.

The decrease of the efficiency is due to the increase of the automatic static common-mode voltage, which induces an increase of the reverse current under the large RF input conditions. Thus, it acts as a self-output power regulation system.

Concerning the dependency of PCE on frequency as illustrated in Figure 1.20, we noted that the PCE is constantly reduced when the frequency is increased. This is due to the AC resistive loss which becomes higher and the detrimental effect of the decrease of the input reactance due to C_{in}.

The efficiency reasonably varies with the load resistance. It changes from 90% for a 100 kΩ load corresponding to an input power of −24 dBm to a 65% for a 5 kΩ load corresponding to an input power of −9 dBm.

In a well-designed rectifier circuit, the performance should not depend strongly on the transistor size. Again, there is a compromise to make between the reduction of the on-resistance and the increase of the input capacitance leading to larger parasitic loss and leakage in the case of large transistors. On the other hand, if the transistors are too small [YI 07], then the charge transfer is incomplete, leading to a low output voltage and thus low efficiency. Hence, there must be an optimal size of transistors to maximize efficiency and output voltage.

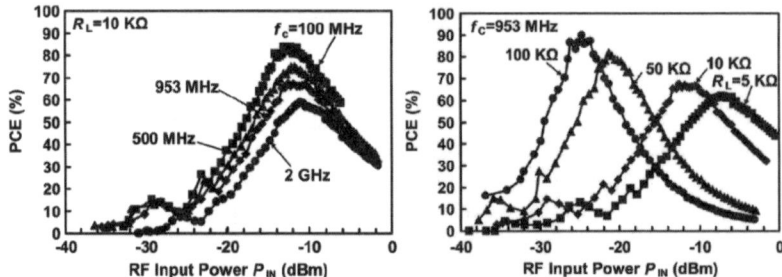

Figure 1.20. *PCE and DC as a function of Pin with frequency and output resistance as parameters (after [KOT 09])*

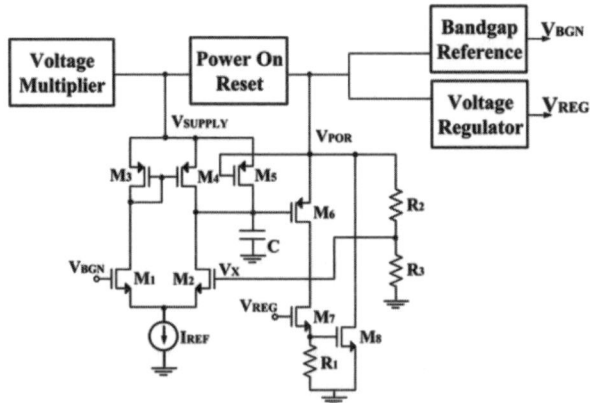

Figure 1.21. *Voltage limiter (after [FER 12])*

1.5. Voltage limiter and regulator

Following the multiplier is the voltage limiter. A voltage limiter is required to avoid damages due to overvoltages whenever the tag and the reader are in close proximity.

The new architecture, which was proposed by Fernandez [FER 12], as compared with the conventional one (chain of diode-connected transistors to ground) corrects two drawbacks: first, the voltage limiter experiences less variations with respect to temperature and process dispersion. The voltage deviation is only ±65 mV compared to ±175 mV. To achieve that, it takes advantage of the power-on reset, bandgap reference and voltage regulator blocks. The current consumption is reported to be only 150 nA when the reader and the tag are far away, thus not degrading the sensitivity.

Following the voltage limiter is the voltage regulator. Due to the high modulation index used to encode the reader commands, it is not obvious to achieve a good line regulation. Besides, the power consumption of this regulator should be minimized to preserve efficiency so as to guarantee tag IC operation. One problem when designing a low-power regulator is to preserve the stability of the closed-loop system. Figure 1.22 illustrates the use of an embedded current with a feed-forward stage buffer compensation feedback stage [LEE 14] to displace the poles according to the current drawn. The circuit provides an output voltage of 1.3 V for an input coarse voltage (limited voltage) of 2.1 V and supplies a load current of 1–60 μA.

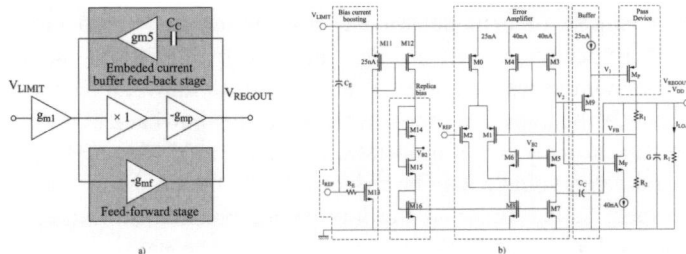

Figure 1.22. *a) Block diagram of the regulator and b) schematic of the regulator(after [LEE 14])*

1.6. Demodulator

Cost prohibits the use of detection technology other than envelope detection for the amplitude modulation (ASK) or compatible schemes (DSB-ASK, SSB-ASK, PR-ASK) specified in EPC Class1Gen2 protocol. Most of the teams worldwide use the historic scheme illustrated in Figure 1.23 and proposed by the German team (Atmel) [KAR 03]. They are all based on the principle of edge detection.

So, this first block is rather similar to the circuit used in the rectifier and has the purpose of extracting the envelope of the input carrier signal. Of prime importance is the value of the capacitor of the envelope detector, together with the current sink, that should be dimensioned to extract the minimum width of gaps (approximately 4 μs) of the incoming signal. Specifications for the data envelope are given in the RFID air interface protocol. This circuit behaves as a low-pass filter that screens out the UHF carrier residue. The signal is then fed into a hysteresis comparator to generate the output data flow. The input signal to the comparator follows the input signal of the tag and displays a large dynamic range. The designer should take into account the average value of this signal to determine the hysteresis window. Then, the signal is integrated and a simple discriminator decides the length of the pulse. The integrator is used to reset by the output of the comparator inverted. This is also used as a system clock and can be reused for the internal oscillator, later used for the modulator.

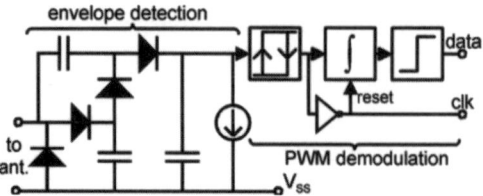

Figure 1.23. *Block diagram of the demodulator (after [KAR 03])*

A new schematic of demodulator has been proposed by Nakamoto (Fujitsu) [NAK 07] that performs well in the case where EEPROM is replaced by ferroelectric random access memory (FeRAM) (which allows a nearly three times faster Read and Write transition time). In this case, the drawback of the conventional voltage detection scheme is that it operates in voltage mode (proportional to power) and the demodulated voltage may be reduced due to the necessary saturation curve (V_{in} with respect to P_{in}) to avoid the low device breakdown voltage of FeRAM (approximately 4 V).

The idea of obtaining a large modulated signal over the entire communication range as illustrated in Figure 1.24 is to convert the ASK input voltage into an ASK input current by maintaining the tag input voltage at a constant value. In this case, the tag input current follows the incoming power modulation linearly.

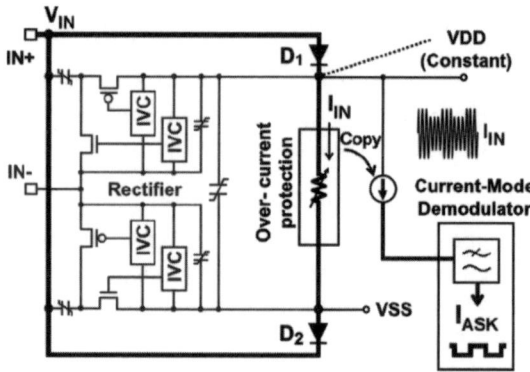

Figure 1.24. *Block diagram of the current-sensing method (after [NAK 07])*

Figure 1.25 shows the block diagram of the current-mode demodulator that outputs the ASK voltage-received data. The core of it is the current comparator that compares the difference between the current-mode ASK raw data minus its peak value with a dynamic threshold, which is the peak value divided by a certain quantity approximately 10% to satisfy the minimum modulation index of 15% [NAK 07]. The authors claim a 27 dB linear dynamic range for the ASK current, corresponding to a communication range from 0 to more than 4 m.

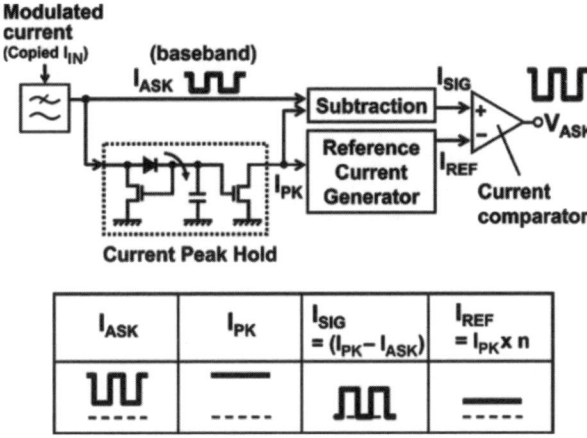

I_{ASK}	I_{PK}	I_{SIG} $= (I_{PK} - I_{ASK})$	I_{REF} $= I_{PK} \times n$
⊓⊔	–––––	⊓⊓	–––––

Figure 1.25. *Block diagram of the current-mode demodulator (after [NAK 07])*

1.7. Oscillator

Backscattering is the principle used in RFID air interface protocol, as will be seen in Chapter 2. An RFID tag needs a clock generation circuit to output the returned data with the proper rate. A clock cannot be extracted from the input signal, which is too high in frequency (prohibitive consumption of the dividers); then, an on-board clock generator has to be designed. Fortunately, the accuracy is not a stringent constraint because protocols define relative timing rather than absolute timing; for example, EPC C1G2 uses an FM0 or Miller encoded subcarrier for the tag to reader link [ZHU 05].

When a communication is established between the tag and the reader, the latter sends a calibration signal to the tag, constituted of a series of eight pulses and each pulse being 116 μs long. At the end of this pulse, there are separation pulses that allow the tag chip to adjust its oscillation frequency as illustrated in Figure 1.26.

If each pulse received by the chip is measured with a counter clocked at 2.2 MHz, then the counted value will be 255 (116 μs × 2.2 MHz). If the frequency of the oscillator varies, the count number will vary accordingly. So, depending on the count number, each bit of successive approximation register (SAR) in the digital control block is set or reset, and the oscillation is adjusted [LEE 09].

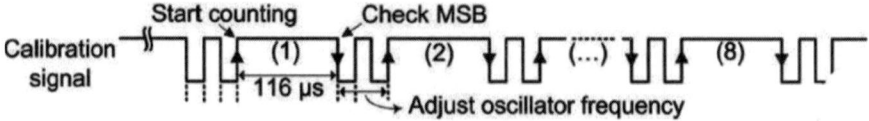

Figure 1.26. *Calibration signal from reader to tag for oscillator adjustment (after [LEE 09])*

Figure 1.27 shows the schematic of a ring-type voltage-controlled oscillator (in competition with the relaxation oscillator due to its low power consumption) with digital calibration, which was proposed by Lee but introduced first by Najafi. This circuit can compensate for ±50% variation of the frequency due to process, voltage and temperature (PVT) variations with a final accuracy of less than 0.5%, satisfying the standard figure (which is less than ±15%).

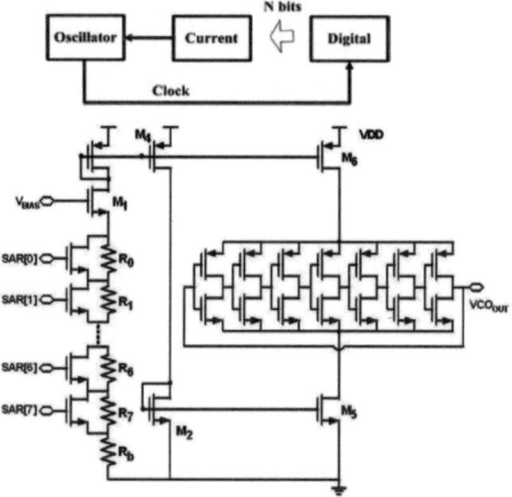

Figure 1.27. *Calibration signal from reader to tag for oscillator adjustment (after [LEE 09])*

1.8. Modulator

In the reverse link, the most suitable modulation scheme is backscattering. Backscattering is a low-power modulation scheme in which the RFID tag reflects a part of the incident RF power back to the reader [ASH 09] by changing the input impedance of the IC.

ASK and PSK are the two modulation formats defined in the EPC Gen2 protocol. In the ASK modulation scheme, the two possible modulation states are obtained by changing a pure resistance where one state is a perfect match ($R_{in} = R_a =$ and the other state is close to short circuit. In the PSK case, only the imaginary component of the switched impedance is changed whereas the resistive part is kept to match with the antenna. The classical way to implement this latter scheme is to modify the reactance of a MOSFET or varactor device. PSK is usually preferred to ASK because of its higher quality of data communication link in terms of bit error rate (BER) and constant power supply to the transponder, but ASK has the advantage of a smaller occupied area and frequency independency [MOH 06]. To illustrate this, we have chosen the phase modulator developed by Karthaus and Fischer [KAR 03].

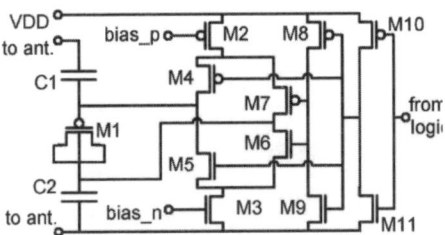

Figure 1.28. *Backscatter phase modulator (after [KAR 03])*

In this circuit, the modulating reactance is achieved using an accumulation mode MOS varactor M1. The DC voltage across the varactor is varied between plus and minus V_{DD}, thereby changing its capacitance between maximum and minimum values.

1.9. Digital blocks

In an RFID tag, less power consumption means the longer communication range. Therefore, it is essential to reduce the consumption of the digital, which corresponds to the major part of the power consumed. Usually, the static part is much lower than the dynamic power consumption. It is given by [KIM 12]:

$$P_{dyn} = 0.5 \cdot \alpha \cdot C_{load} \cdot V_{DD}^2 \cdot f \qquad [1.24]$$

where α is the switching activity (to generate all the necessary clocks), C_{load} is the load capacitance, V_{DD} is the bias voltage and f is the clock frequency (usually approximately 2 MHz).

The designer should not forget the problem of the peak digital power consumption that limits reading sensitivity. Some authors [KIM 12] use adaptive techniques in order to use less power when available power is reduced.

1.9.1. *Memory*

Conventional tag ICs use EEPROM as the rewritable non-volatile memory (NVM). However, the typical 10 times more power required in the case of writing than reading a tag, in the case of EEPROM (charge-pump generator), has led to reduced write range, typically 80%. Researchers [NAK 07] have used FeRAM as the replacement to avoid the charge pump, enabling a nearly three times faster read-and-write transaction time.

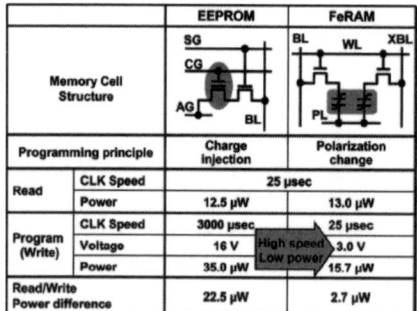

		EEPROM	FeRAM
Memory Cell Structure			
Programming principle		Charge injection	Polarization change
Read	CLK Speed	25 µsec	
	Power	12.5 µW	13.0 µW
Program (Write)	CLK Speed	3000 µsec	25 µsec
	Voltage	16 V	3.0 V High speed / Low power
	Power	35.0 µW	15.7 µW
Read/Write Power difference		22.5 µW	2.7 µW

TABLE III
COMPARISON OF TRANSACTION TIME AND THROUGHPUT FOR TAGS WITH EEPROM AND FeRAM

		Tag with EEPROM	Tag with FeRAM
Inventory (ID Search) Ability		~100 tags/sec	
Command Operation	Read	3.6 msec	
	Write	19.4 msec	4.2 msec
Read/Write Ability		44 tags/sec	129 tags/sec
Write Time Percentage		56.2 %	0.9 %

Figure 1.29. *Comparison of performances between EEPROM and FeRAM (after [NAK 07])*

An alternative is to use an NVM [NAJ 10, KIM 12], which is compatible with a standard CMOS process and also reduces the power consumption drastically.

As illustrated in Figure 1.29, a comparison between FeRAM and EEPROM shows a consumption divided by 2 in the write mode and similar in the read mode. With NVM, another 5 to 10 factor allows drastic improvements.

1.10. Technology, performances and trends

The choice of the technology is essential in the RFID application domain. The cost objective is roughly $0.05 per chip in a small-volume production, which leaves a very small margin to develop the antenna and the bonding process, not to exceed the $0.05 tag cost goal promulgated by organizations such as EPCglobal. On the other hand, the only important technical constraint is the power consumption of the chip, because contrary to technologies used in the digital communication fields such as Wi-Fi, speed is not a real constraint in RFID applications because of the low data rate. To reach this cost by insuring the technical constraint of power consumption, the designer has to tackle different key points.

1.10.1. *Technology choice*

1.10.1.1. *Choice in terms of rectifying device implantation*

For passive UHF RFID tags, low turn-on voltage Schottky diodes or low or zero V_{th} diode-connected MOSFETs are typically used as rectifying devices. In microwave applications, Schottky diodes are usually fabricated in specialized processes where parameters such as barrier height, series resistance or capacitance can be fully controlled. As a result, the Schottky bulk diode (SBD) proves to be superior to the diode-connected MOSFET for the design of rectifiers.

In RFID, low cost pushes toward Schottky diode compatibility in a standard CMOS process. Several research works have been published [ZHU 05] to prove this is somewhat feasible. Due to the low series resistance, small threshold voltage and low junction capacitance, silicon–titanium Schottky diodes can be used in the rectifier.

In terms of cost, it is definitely more expensive to use Schottky diodes due to the mask process even if one reads in some articles that this is possible with no added cost.

Hence, much effort has been made to develop other CMOS solutions to overcome the SBD problem encountered by most CMOS processes. As we have seen in the proposed voltage multiplier circuits, alternate solutions proposed include the use of a programmed threshold by a bias circuitry, the use of an automatic static compensation threshold or the use of an automatic dynamic compensation threshold MOS DTMOST [TEH 09]. This is accomplished by tying the gate and the substrate together (instead of tying the body to the circuit ground). This allows the transistor to have a low leakage current when reversed biased and a high current drive (a lower threshold voltage due to the forward-bias body effect) when turned on.

A Schottky diode exhibits an exponential relationship between its current and voltage, whereas a diode-connected transistor exhibits a square law relationship. The choice of using one or the other depends on the availability of devices, cost, MOS threshold voltage variation and Schottky diode threshold variation.

When choosing the MOSFET, designers may have the choice between transistors featuring different threshold voltages. Low-threshold transistors are preferentially chosen for the front-end rectifier and high threshold transistors are for the digital core part so as to minimize the off-current leakage contribution to the power consumption. Target supply voltage can be regulated close to the transistor threshold (i.e. V_{dd} = 0.6 V) in order to limit the dynamic power consumption. Even lower power consumption could be reached by exploiting the subthreshold device operation. This is possible because it is enough for implementing low data rate RFID protocols. However, the main drawback in this case is the vulnerability of the circuit performance to variations in manufacturing. For instance, the turn-on voltage spread may vary considerably from one chip to the other or small fluctuations on power supply voltage would result in a large spread on critical path delays of digital circuitry [RIC 07].

A shorter MOSFET gate length should improve the performances because of the reduction in the value of the parasitics (e.g. capacitances). But actually, it is becoming less valuable to maintain the effort in trying to reduce the gate length of MOS. This is because we have reached the limits of the scalable model. For example, it is hard to consider the undergate over speed effect to model the leakage current in the ultrathin oxide layer (means a gate parasitic current).

1.10.1.2. *Choice in terms of substrate choice*

Preferably, the substrate should minimize the parasitic capacitances to ground. This will lower the static top plate coupling capacitance and interconnect line capacitance to ground. In this case, the substrate will be of silicon-on-insulator substrate [CUR 05]. This is more or less mandatory if one wants to work in the higher frequencies or super high frequency (SHF).

1.10.2. Design optimization

The hungry block of the analog part is the oscillator as seen before, and the architectural techniques used to reduce the power consumption of the digital block are [NAJ 10]:

– modular design strategy;

– turning off idle modules;

– resource sharing;

– use of NVM register;

– clock data recovery scheme.

At logic synthesis level, clock gating is used to reduce the switching activity of the sequential cells by removing useless edges of the clock signal, for example by limiting the operation speed of the system by the forward or reverse link data rates. Other techniques such as the gate resizing allow a significant reduction in the power consumption. Najafi reports a reduction by 10 (68 to 6 µW) by using clock gating and gate resizing (plus another technique operand isolation that does not contribute much).

1.10.3. *Circuit performances*

Several passive UHF RFID transponders have been published with various sensitivities; however, they often implement a simpler protocol than the one required by the EPC Gen2 specification. Implementing the complete complex protocol induces additional processing power and places a higher constraint on the transponder when we talk about sensitivity, or ultimately on maximum read/write distance. A selection of results is presented in Table 1.2 whenever the information given were more or less complete (the sensitivity is often mentioned for the Read mode but not the Write mode, which is usually more than 5 dB).

Team	Reference	Sens (Read) (dBm)	Consu. (mW) R/W	Volt. (V)	CMOS (mm)	Memory	Die Area (mm²)
Atmel (Germany)	KAR 03	−17	16/35	1.5	0.5	EEPROM	?
Fujitsu (Japan)	NAK 07	−10	80	?	0.35	FeRAM 2 k	1.8
Texas Ins. (USA)	BAR 07	−14	2.7/	1.45	0.13	EEPROM 192 B	0.55
Unistar (Canada)	NAJ 10	−11	12	1	0.18	NVM 256 B	0.95
Hong Kong University	YIN 10	−6	32/?	0.8	0.18	OTP 128 B	1.1
Keti Tech. Inst. (Korea)	KIM 12	−15	6	1.3	0.18	NVM	0.38
Kyung He Univ. (Korea)	LEE 14	−9.2	29/71	1.2	0.13	OTP 4 k	1.1

Table 1.2. *Listing of performances for different design teams*

The performances of the commercially available products are largely as good as the laboratory products with the advantage of being sure that the tag fully satisfies the protocol in perfectly defined test conditions. Figure 1.30 shows the product Ucode7 proposed by NXP as an example.

Symbol	Parameter	Conditions		Min	Typ	Max	Unit
f_i	input frequency			840	-	960	MHz
$P_{i(min)}$	minimum input power	READ sensitivity	[1][3][8]	-	−21	-	dBm
$P_{i(min)}$	minimum input power	WRITE sensitivity	[2]	-	−16	-	dBm
t 16bit	Encoding speed	16-bit	[5]	-	1	-	ms
		32-bit (block write)	[5]	-	1.8	-	ms
C_i	Chip input capacitance	parallel	[3][4]	-	0.63	-	pF
z	Chip impedance	866 MHz	[3][4]	-	14.5-j293	-	Ω
		915 MHz	[3][4]	-	12.5-j277	-	Ω
		953 MHz	[3][4]	-	12.5-j267	-	Ω
z	Typical assembled impedance [9]	915MHz	[6]	-	18-j245	-	Ω
z	Typical assembled impedance [9] in case of single-slit antenna assembly	915MHz	[6][7]	-	13.5-j195	-	Ω
Tag Power Indicator mode							
$P_{i(min)}$	minimum input power level to be able to select the tag		[2]	-	−15	-	dBm

[1] Power to process a QUERY command
[2] Tag sensitivity on a 2dBi gain antenna
[3] Measured with a 50 Ω source impedance directly on the chip
[4] At minimum operating power
[5] When the memory content is "0000...".
[6] The antenna shall be matched to this impedance
[7] Depending on the specific assembly process, sensitivity losses of few tenths of dB might occur
[8] Results in approximately -21,5dBm tag sensitivity with a 2dBi gain antenna
[9] Assuming a 80fF additional input capacitance, 250fF in case of single slit antenna

Figure 1.30. *Performances of the NXP product Ucode7*

1.11. Bibliography

[ASH 07] ASHRY A., SHARAF K., "Ultra low power UHF RFID tag in 0.13μm CMOS", *IEEE International Conference on Microelectronics (IEEE ICM '07)*, pp. 109–126, December 2007.

[ASH 09] ASHRY A., SHARAF K., IBRAHIM M., "A compact low-power UHF RFID tag", *Microelectronics Journal*, vol. 40, pp. 1504–1513, 2009.

[BAG 09] BAGHAEI M., ZOU Z., MENDOZA D.S., *et al.*, "Remotely UHF-powered ultra wideband RFID for ubiquitous wireless identification and sensing", in TURCU C. (ed.), *Development and Implementation of RFID Technology*, Open Access Database, pp. 109–126, February 2009.

[BAL 06] BALACHANDRAN G., BARNETT R., "A 110nV voltage regulator system with dynamic bandwidth boosting for RFID systems", *IEEE Journal of Solid-State Circuits*, vol. 41, no. 9, pp. 2019–2028, September 2006.

[BAL 09] BALACHANDRAN G., BARNETT R., "A high dynamic range ASK demodulator for passive UHF RFID with automatic over-voltage protection and detection threshold adjustment", *IEEE Custom Integrated Circuits Conference (CICC '09)*, pp. 383–386, 2009.

[BAR 06] BARNETT R., LAZAR S., LIU J., "Design of multistage rectifiers with low-cost impedance matching for passive RFID tags", *Digest IEEE Radio Frequency Integrated Circuits*, pp. 291–294, June 2006.

[BAR 07] BARNETT R., BALACHANDRAN G., LAZAR S., *et al.*, "A passive UHF RFID transponder for EPC Gen2 with -14 dBm sensitivity in 130 nm CMOS", *Digest of Technical Papers, 2007 IEEE International Solid-State Circuits Conference (ISSCC)*, pp. 582–583, 11–15 February 2007.

[BAR 08] BARNET R., LIU J., "An EEPROM programming controller for passive UHF RFID transponders with gated clock regulation loop and current surge control", *IEEE Journal of Solid-State Circuits*, vol. 43, no. 8, pp. 1808–1815, August 2008.

[BAR 09] BARNETT R., LIU J., LAZAR S., "A RF to DC voltage conversion model for multi-stage rectifiers in UHF RFID transponders", *IEEE Journal of Solid-State Circuits*, vol. 44, no. 2, pp. 354–370, February 2009.

[CHU 11] CHUNG C., KIM Y., KI T., *et al.*, "Fully integrated ultra-low-passive UHF RFID transponder IC", *IEEE International Symposium on Radio-Frequency Integration Technology*, Beijing, China, pp. 77–80, 30 November – 2 December 2011.

[CUR 05] CURTY J.P., JOELH N., DEHOLLAIN C., *et al.*, "2.45 GHz remotely powered RFID system", *Proceedings of the Conference on PhD Research in Microelectronics and Electronics 2005*, vol. 1, pp. 127–130, 2005.

[CUR 07] CURTY J.P., DECLERCQ M., DEHOLLAIN C., *et al.*, *Design and Optimization of Passive UHF RFID Systems*, Chapters 2.1, 2.2, 2.3, 5 and 7, 1st ed., Springer, 2007.

[DOB 07] DOBKIN D.M., *The RF in RFID: Passive UHF RFID in Practice*, Chapter 5, Newnes, Burlington, MA, 2007.

[DON 10] DONG-SHENG L., XUE-CHENG Z., KUI D., *et al.*, "New design of RF rectifier for passive UHF RFID transponders", *Microelectronics Journal*, vol. 41, pp. 51–55, 2010.

[DU 12] DU Y., LI X., DAI L., *et al.*, "A new type of low-power read circuit in EEPROM for UHF RFID", *Microelectronics Journal*, vol. 43, pp. 364–369, 2012.

[FER 12] FERNANDEZ E., BERIAIN A., SOLAR H., *et al.*, "A low-power voltage limiter for a full passive UHF RFID sensor on a 0.35µm CMOS process", *Microelectronics Journal*, vol. 43, pp. 708–713, 2012.

[GUO 10] GUO J., SHI W., LEUNG K., *et al.*, "Power-on-reset circuit with power-off auto-discharging path for passive RFID tag ICs", *53rd IEEE International Midwest Symposium on Circuits and Systems (MWSCAS '10)*, pp. 21–24, 2010.

[HAS 09] HASHEMI S., SAWAN M., SAVARIA Y., "A novel low-drop CMOS active rectifier for RF-powered devices: experimental results", *Microelectronics Journal*, vol. 40, pp. 1547–1554, 2009.

[HON 09] HONG Y., NG Y.S., CHAN C.F., *et al.*, "A passive RFID tag IC development platform", *3rd International Conference on Anti-Counterfeiting, Security and Identification in Communications (ASID '09)*, 2009.

[JAM 06a] JAMALI B., RANASINGHE D.C., COLE P.H., "Analysis of a UHF RFIC CMOS rectifier structure and input impedance characteristics", *Microelectronics: Design, Technology, and Packaging II, Proceedings of SPIE – the International Society for Optical Engineering*, vol. 6035, 2006.

[JAM 06b] JAMALI B., RANASINGHE D.C., COLE P.H., "Design and optiisation of power rectifiers for passive RFID systems in monolithic circuit", *Microelectronics: Design, Technology, and Packaging II, Proceedings of SPIE – the International Society for Optical Engineering*, vol. 6035, 2006.

[JEO 05] JEON J., "CMOS passive RFID transponder with read-only memory for low cost fabrication", *Proceedings of IEEE International SOC Conference*, pp. 181–184, 2005.

[JIA 10] JIA J., LEUNG K.N., "Optimization of output voltage and stage number of UHF RFID power rectifier", *Proceedings 10th IEEE International Conference on Solid-State and Integrated Circuit Technology (ICSICT '10)*, pp. 412–414, 2010.

[KAR 03] KARTHAUS U., FISCHER M., "Fully integrated passive UHF RFID transponder IC with 16.7µW minimum RF input power", *IEEE Journal of Solid-State Circuits*, vol. 38, no. 10, pp. 1602–1608, October 2003.

[KIM 12] KIM Y.H., KI T.H., CHUNG C., *et al.*, "Implementation of a low-cost and low-power batteryless transceiver SoC for UHF RFID and wireless power transfer system", *Proceedings of the 42nd European Microwave Conference*, Amsterdam, pp. 514–517, 29 October–01 November 2012.

[KOT 09] KOTANI K., SASAKI A., ITO T., "High-efficiency differential-drive CMOS rectifier for UHF RFIDs", *IEEE Journal of Solid-State Circuits*, vol. 44, no. 11, pp. 3011–3018, November 2009.

[LEE 09] LEE J.-W., LEE B., "A long-range UHF-band passive RFID tag IC based on high-Q design approach", *IEEE Transactions on Industrial Electronics*, vol. 56, no. 7, pp. 2308–2316, July 2009.

[LEE 10] LEE K.-S., CHUN J., KWON K., "A low power CMOS compatible embedded EEPROM for passive RFID tag", *Microelectronics Journal*, vol. 41, pp. 662–668, 2010.

[LEE 14] LEE J.-W., PHAN N.D., THAI Vo D.H., *et al.*, "A fully integrated EPC-Gen2 UHF-band passive tag IC using an efficient power management technique", *IEEE Transactions on Industrial Electronics*, vol. 61, no. 6, pp. 2922–2932, 2014.

[LOO 08] LOO C.H., ELMAHGOUB K., YANG F., *et al.*, "Chip impedance matching for UHF RFID tag antenna design", *Progress in Electromagnetics Research*, vol. 81, pp. 359–370, 2008.

[MAN 07] MANDAL S., SARPESHKAR R., "Low-power CMOS rectifier design for RFID applications", *IEEE Transactions on Circuits and Systems-I*, vol. 54, no. 6, pp. 1177–1188, June 2007.

[MOH 06] MOHD-YASIN F., TEH Y.K., REAZ M.B.I., "Developing designs for RFID transponders", *Microwaves and RF*, pp. 57–86, September 2006.

[NAJ 10] NAJAFI V., MOHAMMADI S., ROOSTAIE V., *et al.*, "A dual-mode UHF EPC Gen2 RFID tag in 0.18μm CMOS", *Microelectronics Journal*, vol. 41, pp. 458–464, 2010.

[NAK 06] NAKAMOTO H., YAMAZAKI D., YAMAMOTO T., *et al.*, "A passive UHF RFID tag LSI with 36.6% efficiency CMOS-only rectifier and current-mode demodulator in 0.35μm FeRAM technology", *IEEE International Solid-State Circuits Conference, Digest of Technical Papers*, pp. 307–310, 2006.

[NAK 07] NAKAMOTO H., YAMAZAKI D., YAMAMOTO T., *et al.*, "A passive UHF RF identification CMOS tag IC using ferroelectric RAM in 0.35μm technology", *IEEE Journal of Solid-State Circuits*, vol. 42, no. 1, pp. 101–110, January 2007.

[RIC 07] RICCI A., DE MUNARI I., "Enabling pervasive sensing with RFID: an ultra low-power digital core for UHF transponders", *Proceedings of IEEE International Symposium on Circuits and Systems*, pp. 1589–1592, 2007.

[RON 06] RONGSAWAT K., THANACHAYANONT A., "Ultra low power analog front-end for UHF RFID transponder", *2006 International Symposium on Communications and Information Technologies (ISCIT)*, pp. 1195–1198, 2006.

[SAS 08] SASAKI A., KOTANI K., ITO T., "Differential-drive CMOS rectifier for UHF RFIDs with 66% PCE at –12 dBm input", *Proceedings of the 2008 IEEE Asian Solid-State Circuits Conference (ASSCC '08)*, pp. 105–108, November 2008.

[SHE 13] SHEN J., WANG X., WANG B., *et al.*, "Fully integrated passive UHF RFID transponder IC with a sensitivity of -12 dBm", *Proceedings of IEEE International Symposium on Circuits and Systems (ISCAS '13)*, pp. 289–292, 2013.

[TEH 09] TEH Y.K., MOHD-YASIN F., CHOONG F., *et al.*, "Design and analysis of UHF micropower CMOS DTMOST rectifiers", *IEEE Transactions on Circuits and Systems-II*, vol. 56, no. 2, pp. 122–126, February 2009.

[UME 06] UMEDA T., YOSHIDA H., SEKINE S., *et al.*, "A 950-MHz rectifier circuit for sensor networks tags with 10-m distance", *IEEE Journal of Solid-State Circuits*, vol. 41, no. 1, pp. 35–41, January 2006.

[VIT 05] DE VITA G., IANNACONE G., "Design criteria for the RF section of UHF and microwave passive RFID transponders", *IEEE Transactions on Microwave Theory and Techniques*, vol. 53, no. 9, pp. 2978–2990, September 2005.

[WON 11] WONG S.-Y., CHEN C., "Power efficient multi-stage CMOS rectifier design for UHF RFID tags", *Integration, the VLSI Journal*, vol. 44, pp. 242–255, 2011.

[YAO 05] YAO Y., SHI Y., DAI F., "A novel low-power input-independent MOS AC/DC charge pump", *IEEE International Symposium on Circuits and Systems 2005*, vol. 1, pp. 380–383, 23–26 May 2005.

[YI 07] YI J., KI W., TSUI C., "Analysis and design strategy of UHF micro-power CMOS rectifiers for micro-sensor and RFID applications", *IEEE Transactions on Circuits and Systems – I*, vol. 54, no. 1, pp. 153–166, January 2007.

[YIN 10] YIN, J., YI J., LAW M., *et al.*, "A system on-chip EPC Gen2 passive UHF RFID tag with embedded temperature sensor", *IEEE Journal* of *Solid-State Circuits*, vol. 45, no. 11, pp. 2404–2420, November 2010.

[ZHA 09] ZHANG S., SUN L., HONG H., *et al.*, "A low-power analog front end of passive UHF RFID tag IC for EPC^{TM}C1C2", *Proceedings of 8th IEEE International Conference on ASIC (ASICON '09)*, pp. 557–560, 2009.

[ZHU 04] ZHU Z., JAMALI B., COLE P., Brief comparison of different rectifier structures for RFID transponders, Report of Auto-ID Lab, University of Adelaide, 2004.

[ZHU 05] ZHU Z., JAMALI B., COLE P.H., "An HF/UHF RFID analogue front-end and analysis", *White paper series*, 1st ed., Auto-ID Lab, September 2005.

2

Design of UHF RFID Tags

2.1. Tag antenna design

Radio frequency identification (RFID) tag antennas present unique design challenges to satisfy heavy constraints due to cost and size. The antenna miniaturization on inexpensive materials is only one of several problems that a designer needs to solve, others being the wideband impedance matching to the RFID integrated circuit (IC) and the sensitivity to the environment. Section 2.1.1 describes the fundamental circuit parameters of the dipole antenna. As most ultra high frequency (UHF) RFID tags are dipole-based structures, dipole issues regarding the input impedance, radiation resistance, efficiency, quality factor (Q-factor) and impedance will be of practical interest in the design of tag antennas. For polarization, radiation pattern and propagation issues, readers can refer to [BAL 05] and [DOB 12]. Then, miniaturization strategies based on fat dipoles, tip loading and meanders are presented in sections 2.1.2 and 2.1.3. The last section (section 2.1.4) addresses the influence of the dielectric and metallic environments on the tag performance.

2.1.1. *Fundamental circuit parameters of the dipole antenna*

2.1.1.1. *Equivalent electrical circuit*

The dipole forms the basis of most tag antenna designs. It can be built with either metallic wires or strips. Considering the dipole in Figure 2.1, when a potential is applied to the antenna inputs, there is opposite charge buildup on the ends. Essentially, the dipole ends can be viewed as open circuits, with a high voltage and no current. Due to the charge buildup at either end of the dipole, current begins to flow. Therefore, any antenna simultaneously stores charge and opposes changes in current. These phenomena are, respectively, associated with a capacitor and an inductor in an electrical model. If the distance between positive and negative sources is comparable with a wavelength, the retardation in going from different parts of the dipole to a point in space prevents radiation cancellation and produces reinforcement of effects in a certain direction. This energy flow into the environment can be modeled by a resistance.

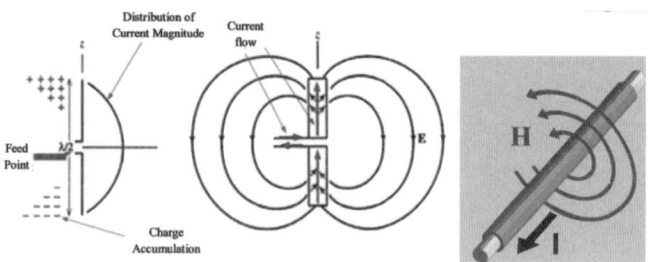

Figure 2.1. *Charge and current distributions in a dipole antenna. Near-field distributions of the E- and H-components*

Ignoring the influence of the antenna thickness, the length of a UHF band half-wave dipole antenna is about 16.4 cm at 915 MHz. Over a relatively narrow bandwidth (5%) near resonance, the half-wave dipole resonance can be fairly well predicted by a series RLC circuit (Figure 2.2). The reactance tends to follow the series LC circuit but the resistance is not constant in reality. The resonant frequency of such a circuit is given by:

$$f_0 = \frac{1}{2\pi\sqrt{LC}}$$
[2.1]

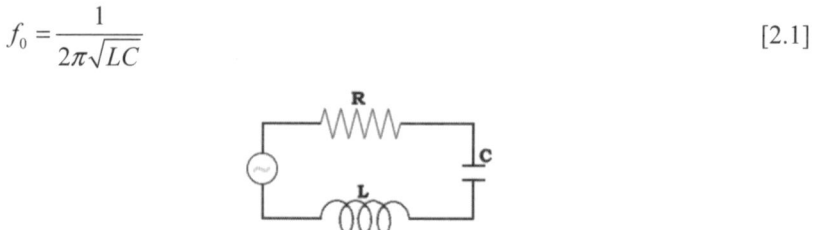

Figure 2.2. *Equivalent series RLC circuit of a half-wave dipole antenna*

The basic half-wave dipole is not the appropriate solution in UHF RFID for two reasons. First, its length is too long for UHF labels. Second, the antenna reactance at resonance is too low and its resistance is too high to match the typically low equivalent series resistance and high capacitance of UHF RFID ICs.

Antennas in the RFID industry often have a length of approximately 92 mm because of the convenience of placing tags within a 102.6 mm wide label. But a straight dipole shorter than a half-wave has a negative reactance. Therefore, an additional inductance must be synthesized to make the short antenna resonate at a lower frequency than its natural resonant frequency. At this stage, it is useful to introduce the following relationships [MCD 12] for the total capacitance C and inductance L of a short wire dipole of length l and diameter b:

$$C \approx \frac{\pi \varepsilon_0 l}{2 \ln(l/b)}$$ [2.2]

$$L \approx \frac{\mu_0 l}{4\pi} \ln(l/b)$$ [2.3]

These equations are valid for $l \ll c/f_0 = \lambda_0$. For $l = 3.33$ cm, i.e. $l = \lambda_0/10 \ll \lambda_0$ at 900 MHz, we would find $C = 0.26$ pF and $L = 11.6$ nH with $b = 1$ mm. It is easy to show that the reactance $L\omega - 1/C\omega$ of a short linear antenna is largely due to its capacitance. By replacing the rough estimates of L and C in [2.1], we obtain $l = \lambda_0/2.22$, which gives a fairly good prediction of "resonance" and physical modeling even for a half-wave dipole.

2.1.1.2. Radiation resistance, input impedance, efficiency and gain

The radiation resistance R_{rad} of an antenna is defined as the resistance that would dissipate the same amount of power as the antenna radiates, when the current in this resistance is equal to the current maximum in the antenna. On the other hand, the input impedance Z_a of any antenna is defined as the ratio of the voltage and the current at its terminals. For a center-fed dipole, the maximum current and the current at the terminals only coincide if $l \leq \lambda_0$ (Figure 2.3). Under these conditions, we can state that the input resistance equals the radiation resistance ($R_a = \text{Re}(Z_a) = R_{rad}$) for a lossless dipole.

The input impedance Z_{as} of a dipole with an offset feeding (asymmetric) can be related to the input impedance Z_s of a center-fed (symmetric) dipole of the same length. The impedances Z_{as} and Z_s relate to the radiated power through their respective current magnitudes at the dipole terminals I_{as} and I_s (Figure 2.4). Imposing the condition that, for a given current distribution, the power delivered by the transmitter must be equal to the sum of the radiated (active) and stored (reactive) power of the antenna leads to

$$Z_s I_s^2 = Z_{as} I_{as}^2$$ [2.4]

The offset feeding of the asymmetric dipole is located at a distance h from the dipole center. Assuming a sinusoidal current distribution, [2.4] becomes:

$$Z_{as}(h) = Z_s \left(\frac{I_s}{I_{as}}\right)^2 = Z_s \frac{\sin^2(kl/2)}{\sin^2\left[k\left(\frac{l}{2} - h\right)\right]}$$ [2.5]

which further reduces to the following expression for a half-wave dipole:

$$Z_{as}(h) = \frac{Z_s}{\cos^2(kh)}$$ [2.6]

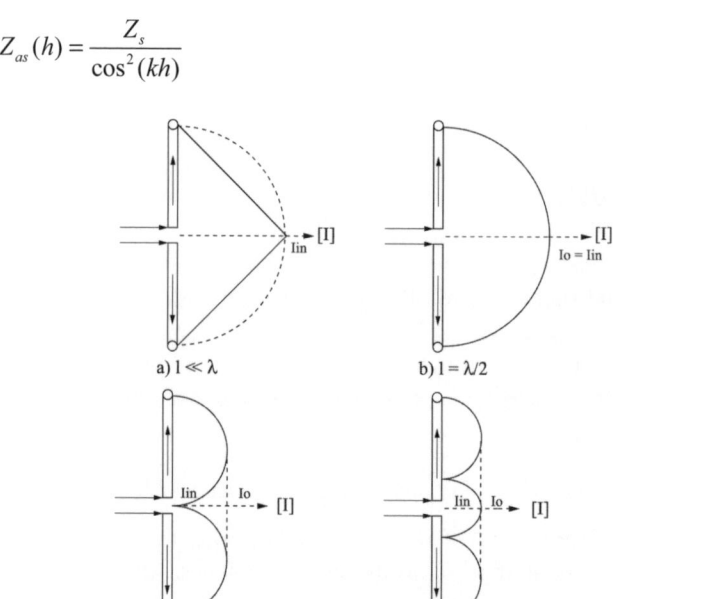

a) $l \ll \lambda$ b) $l = \lambda/2$

c) $\lambda/2 < l < \lambda$ d) $\lambda < l < 3\lambda/2$

Figure 2.3. *Current distribution along the dipole antenna for various antenna lengths*

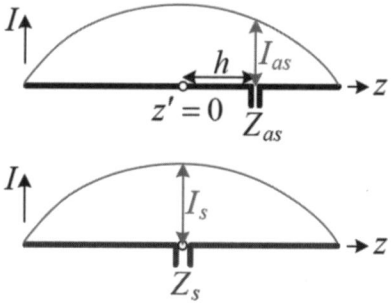

Figure 2.4. *Illustration of the relation between the input impedances for the asymmetric and symmetric dipole feeds*

We conclude that both the input resistance and reactance can be increased by offsetting the antenna input toward the dipole ends where the current goes to 0. It is important to note that the radiation resistance is identical for all positions but it is only in the center-fed case that the real part of the input impedance equates the radiation resistance.

In a lossy antenna, the input resistance R_a can be separated into a series circuit of two different resistors:

$$R_a = R_{rad} + R_{loss} \qquad [2.7]$$

where R_{rad} is the radiation resistance and R_{loss} is the loss resistance representing the unwanted losses caused by the non-perfect conductors and dielectric materials. The radiation resistance of a wire or strip dipole strongly depends on the current distribution. For the "ideal" ($l \ll \lambda$) dipole, the current is uniform and the radiation resistance is given by [BAL 05]:

$$R_{rad} = \frac{2\pi\eta_0}{3}\left(\frac{l}{\lambda}\right)^2 = 80\pi^2\left(\frac{l}{\lambda}\right)^2 \qquad [2.8]$$

It can be shown that for a small dipole ($\lambda/50 < l \leq \lambda/10$), the current is triangular, maximum at the center and zero at the ends (Figure 2.3). Then, a good approximation is of one-fourth of [2.8]:

$$R_{rad} = 20\pi^2\left(\frac{l}{\lambda}\right)^2 \qquad [2.9]$$

For longer dipoles, the current is distributed as a sine wave. The radiation resistance of a center-fed dipole can be approximated with a high degree of accuracy [TAI 07] as:

$$R_{rad} = -0.4787 + 23.01\,(l/\lambda) + 3.91(l/\lambda)^2 + 484.10(l/\lambda)^3 \qquad [2.10]$$

for a frequency range verifying $f_0/20 < f < 1.2f_0$. Therefore, [2.10] is valid for dipole lengths around a half-wave, the upper limit being $l = 0.6\lambda_0$.

It can be noted that all previous R_{rad} expressions do not depend on the antenna diameter, which is correct as long as l/b is very large. As a result, R_{rad} decreases slightly when the dipole is made thicker. For $b/\lambda = 0.003$ corresponding to $b = 1$mm at 900 MHz, $R_{rad} \sim 62\ \Omega$ whereas the ideal value when $b \to 0$ should be 73 Ω. Although the radius of the wire does not strongly influence the resistances, the gap spacing at the feed plays a significant role, especially when the current at and near the feed point is small.

Finally, the loss resistance due to conductive losses can be calculated with:

$$R_{loss,ohmic} = \sqrt{\frac{\omega\mu_0}{2\sigma}}\,\frac{l}{\pi b} \qquad [2.11]$$

The antenna efficiency η is defined as the ratio:

$$\eta = \frac{R_{rad}}{R_{rad} + R_{loss}} \qquad\qquad [2.12]$$

which is the ratio of the radiated power and the input power of the antenna. Assuming a working frequency of 900 MHz ($\lambda_0 = 33$ cm) and a dipole in free space (no substrate losses), a straight dipole of conductivity $\sigma = 5.7 \times 10^7$ S/m (copper) with $b = 1$mm will be characterized by:

- $R_{rad} = 7.9\ \Omega$, $R_{loss} = 26$ mΩ and $\eta = 99.7\%$ if $l = 3.33$ cm ($l = \lambda_0/10$)

- $R_{rad} = 712$ mΩ, $R_{loss} = 7.8$ mΩ and $\eta = 98.9\%$ if $l = 1$cm ($l = \lambda_0/33$)

To have a better control over the antenna resistance, a loading bar with the same width as the meander trace is added in [RAO 05]. Working as a shunt capacitance, it helps to trim the radiation resistance to the IC input resistance (Figure 2.5).

Figure 2.5. *Meandered dipole with a loading bar for radiation resistance trimming*

The term "antenna gain" describes how much power is transmitted in the direction of peak radiation to that of an isotropic source. Let us assume that an antenna is made up of with conductive materials of zero resistivity and that the dielectrics present no losses. Calling $G_{lossless}$ the gain of this ideal lossless antenna, the peak gain G of the same antenna made up of lossy material is given by:

$$G = \eta G_{lossless} \qquad\qquad [2.13]$$

where η is the efficiency of the lossy antenna. Since antenna gain affects both the efficiency of the power transfer to the chip and the portion of the incident power that will be reflected back to the reader, it is very important to maximize the generally low, due to size, antenna gain.

2.1.1.3. *VSWR bandwidth and Q-factor*

The antenna bandwidth (BW) can be defined from any of its fundamental characteristics such as the return loss, input impedance, polarization and radiation efficiency. In order to relate the antenna bandwidth to the Q-factor, it is more convenient to use the matched voltage-standing-wave-ratio (VSWR) bandwidth

BW_{VSWR}. For an antenna tuned, i.e. showing a zero reactance, at a frequency ω_0, BW_{VSWR} is the difference $\Delta\omega$ between the two frequencies on either side of ω_0 at which VSWR equals a constant S. The fractional matched VSWR bandwidth is then defined as:

$$FBW_{VSWR}(\omega_0) = \frac{\Delta\omega}{\omega_0}$$ [2.14]

This definition is only valid if the characteristic impedance of the feedline equals $Z_a(\omega_0)=R_a(\omega_0)$. The half-power VSWR bandwidth corresponds to $S = 5.828$. The associated value of the magnitude squared of the reflection coefficient is $\alpha = (S-1)^2/(S+1)^2 = 1/2$, which means that one half of the incident power is reflected back.

In the context of resonators, the Q-factor is 2π times the ratio of the energy stored in the resonator (sum of energies stored in lossless inductors and capacitors) to the lost energy per cycle dissipated in resistors at the resonant frequency:

$$Q = 2\pi \times \frac{\text{Energy stored}}{\text{Energy dissipated per cycle}}$$ [2.15]

The concept of the Q-factor is used to describe the antenna as a resonator. Usually, in circuit design we want elements to have a high Q-factor in order to reduce the circuit loss. However, when talking about antennas we want a low Q-factor because the "loss" involved is the radiation we really want. A low-Q antenna shows a wider bandwidth as:

$$Q \approx \frac{1}{\text{Bandwidth}}$$ [2.16]

Conversely, a large-Q antenna is characterized by a sharp resonance and a narrow bandwidth because it stores a lot of energy and radiates relatively little of it.

The concept of the Q-factor is very useful when considering small antennas. The Q-value of the small antenna is high due to the low radiation resistance and the high reactance. The smaller the antenna, the higher the Q-value expected. Hence, the bandwidth of a small antenna is inherently narrow. But the Q-factor can be improved, i.e. reduced for a given antenna volume if appropriate design strategies are applied [BES 05]. A reliable relationship between FBW_{VSWR} and Q has been established in [YAG 05]:

$$Q = \frac{(S-1)/\sqrt{S}}{FBW_{VSWR}(\omega_0)}$$ [2.17]

Figure 2.6 shows the Q-factor as a function of the wire radius $a = b/2$ normalized to the wavelength for a half-wave dipole. It can be observed that the Q-factor is roughly $Q = 4.4$ for a 1.65 mm radius, i.e. $100 \times a/\lambda = 0.5$. However, according to [STU 12, Figure 6.7, pp. 159], FBW_{VSWR} is 16% for $S = 2$ and $100 \times a/\lambda = 0.5$. By replacing these two values in [2.17], we obtain $Q = 4.41$, which is in close agreement with the graphic value.

The worldwide frequency range of UHF RFID tags is defined between 860 and 960 MHz over an 11% fractional bandwidth. We conclude that there is no problem in covering the 11% UHF RFID world band with a straight half-wave dipole. But size reduction techniques described in the next sections (meanders and capacitive loading) increase the Q-factor up to 15, which makes the Q improvement a primary performance issue. Obviously, efforts on the Q-factor could be minimized if the tags are dedicated to a single band, for instance United States (US) only.

An additional drawback of narrowband antennas is that they are more difficult to match and more susceptible to detuning than wideband antennas.

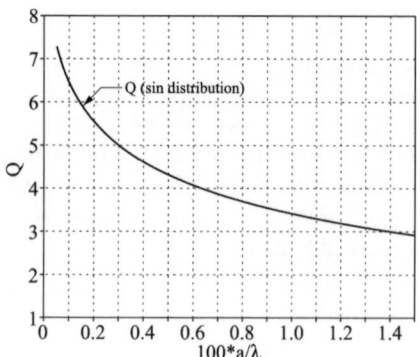

Figure 2.6. *Q factor as a function of dipole radius $a=b/2$ [HAZ 11]*

From Figure 2.6, it is also clear that Q decreases logarithmically with (a/λ). As a result, dividing Q by only 2, for instance from 6 to 3, requires a wire diameter 15 times thicker (from 0.6 mm to 10 mm). So, large variations of the aspect ratio length/diameter result in a weak Q improvement.

[MCL 96] has described the fundamental theoretical limit for the minimum Q-value of a small antenna. If the antenna can be placed inside a bounding sphere of radius r, the minimum Q-value for a lossless antenna is

$$Q_{min} = \frac{1}{(kr)^3} \pm \frac{1}{kr} \qquad\qquad [2.18]$$

where $k = 2\pi/\lambda$. This expresses the absolute minimum Q value the antenna can take. Unfortunately, the theory does not tell us how to implement a minimum Q antenna. However, a key point is that an antenna that fills the volume efficiently will be able to reduce the total reactive fields, resulting in a lower Q. Best [BES 05] has shown that an antenna made up of multiple folded arms in the spherical helix geometry occupies the full spherical volume, achieves self-resonance and an impedance match. The optimal result is within 1.5 times the theoretical lower bound on Q for an electric dipole showing an efficiency of 98%.

But RFID antennas are planar and a dipole only fills a small part of the volume of the bounding sphere. Consequently, the optimal Q-factor which is roughly equal to 1 for a sphere of radius $r = \lambda/2$ using [2.18] is also much lower than the Q values extracted from Figure 2.6.

Up to this point, we have only considered wire dipoles. It useful to admit that the bandwidth of a flat dipole with width w is roughly equivalent to the bandwidth of a wire dipole with diameter b if $w = 2b$ [DEA 10].

2.1.2. Fat antennas and tip loading

As shown in [2.1], at a given resonance frequency, the product of L and C will result in a certain fixed value. Although the product is fixed, the relation of L/C can be chosen arbitrarily. As an example, doubling C and halving L will result in the same resonance frequency. This gives some freedom in the design of the dipole, which means that it is possible to design various kinds of dipoles with different shapes, which all have the same resonance frequency. The difference between two dipoles with the same resonant frequency but a different L/C-ratio is the Q-factor and hence the bandwidth, since

$$Q = \frac{1}{R}\sqrt{\frac{L}{C}} \qquad\qquad [2.19]$$

Clearly, the L/C-ratio has to be minimized in order to obtain a high bandwidth of the dipole. Using the expressions [2.2] and [2.3] for L and C valid for short dipoles, we found that:

$$\sqrt{L/C} \approx \ln(l/b) = -\ln(b/l) \qquad\qquad [2.20]$$

where l and b are the length and diameter of the wire, respectively. We conclude that increasing the dipole diameter favors the Q-factor in two ways: increasing the wire thickness reduces its self-inductance and increasing the conducting surface results in a larger capacitance. [2.10] explains the logarithm dependence of the Q-curve in Figure 2.6.

Broadband fat tags may also increase manufacturing cost, depending on the techniques used for antenna definition. Etching based on subtractive processes makes more sense than expensive silver ink printing if the purpose is to keep maximum of metal. Nevertheless, fat tag antennas have seen wide commercial deployment. Some typical broadband structures are given in Figure 2.7. Manufacturers advise us to use fat tags for high dielectric mounting glass, wood and plastic or challenging metal/plastic/fluid containers.

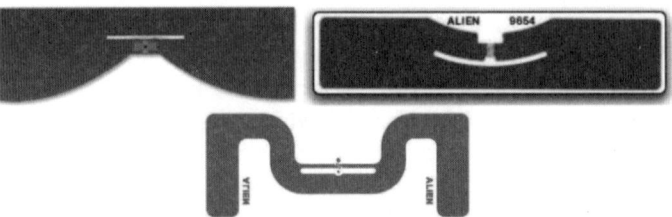

Figure 2.7. *Examples of Alien fat tags with broadband features*

On the other hand, Q is also improved if R is made larger. Obviously, the antenna Q can be reduced by introducing loss in addition to the radiation resistance, but this would reduce the antenna efficiency. We have seen in section 2.1.1.2 that the radiation resistance is basically a function of the wire length. Therefore, increasing l would be a solution but the size reduction is mandatory. The question is: how can we increase R_{rad} with a reduced antenna length? The answer is in the current dependence of R_{rad}. Assuming a dipole length $l = 100$ mm (i.e. $\lambda/3$ at 900 MHz), [2.9] gives $R_{rad} = 21.9$ Ω for a triangular distribution of the current and [2.10] gives $R_{rad} = 24.5$ Ω for a sine distribution. As R_{rad} is four times greater for a uniform current than a triangular one, we would expect 4×21.9 $\Omega = 87.6$ Ω if a uniform distribution of current could be enforced along the wire.

The following question is how can we make the current distribution as uniform as possible when l and λ are of the same order? The idea is:

– to have a straight section at the middle of the dipole where the currents are high. This section is the useful radiating part of the tag;

– to miniaturize the rest of the antenna (dipole ends) in such a way that the antenna resonates by using non-radiating currents. As currents are weaker at the ends of the resonating dipole than in its center, resulting ohmic losses are kept low.

There are basically two practical ways to achieve that:

– to push densely packed meanders at the end of the dipoles. The length extension is used to obtain the resonance (Figure 2.8);

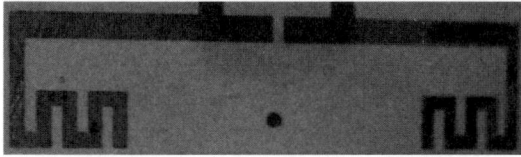

Figure 2.8. *Size reduction by loading the dipole ends with packed meanders*

– to capacitively load the ends of the dipole with large metalized areas storing more electric charges. The resulting capacitance is roughly proportional to the perimeter of the loading shape. The added capacitance compensates the reduced inductance resulting from the shorter dipole length (Figure 2.9) in order to maintain the same resonant frequency. This technique is known as capacitive tip-loading. The capacitor can take the form of a bulk square (Figure 2.10(a)), but flaring the ends of the antenna out to a larger structure (Figure 2.10(b)) and using line reversal (Figure 2.10(c)) are other options.

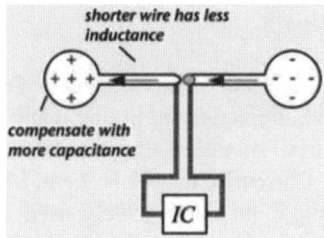

Figure 2.9. *Tip-loading: principle of dipole miniaturization with uniform current [DOB 12]*

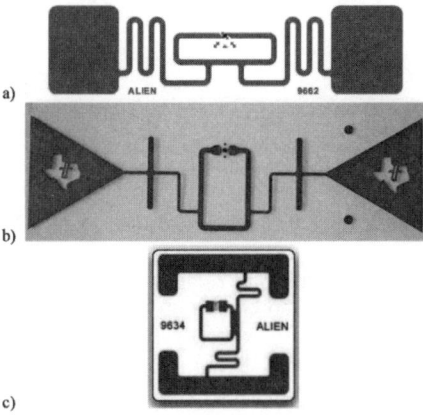

Figure 2.10. *EPC Gen 2 RFID inlays with tip-loadings: a) Alien ALN-9662, b) Texas instruments and c) Alien 9634*

We conclude that capacitive loading has the advantage of physically shortening the element length at the end of the dipole where the current is lowest (least radiation), and without introducing noticeable losses as inductors do. End-loaded short dipoles have the highest radiation resistance, and the intrinsic losses of the loading device are negligible.

2.1.3. Meandered dipoles

In this section, it is shown that a half-wave dipole of length l can be made shorter by folding the wire back and forth and creating meanders (Figure 2.11). For the same resonant frequency as the original half-wave dipole, the resulting meandered dipole antenna (MDA) is characterized by a "mechanical" length s shorter than l in the original horizontal direction without considering the vertical segments. However, the "physical length" $s_{tot}>l$ represents the total wire length of the MDA accounting both for the vertical and horizontal segments. Conversely, the resonant frequency can be decreased by loading a meander line structure onto a half-wave dipole for a fixed antenna length.

Wire-based MDAs can be found in few use cases. For instance, inox cylindrical wires are used in RFID labels implemented in linens and clothes to support the harsh washing process of industrial laundries. But nearly all RFID antennas are some variants of printed MDAs. Currently, copper is the most commonly used conductor in tag antennas and etching is the most widely used manufacturing technique to produce the conductive patterns. However, the tag cost is a crucial factor in the mass production of antennas. Economical manufacturing methods can be achieved by applying printing techniques using conductive ink. In printed electronics, silver particles are often used to form the conductive layer of the metal-line on the tag antenna.

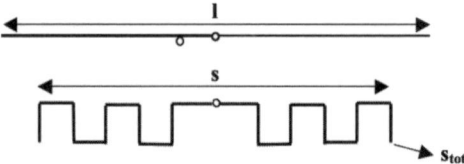

Figure 2.11. *Typical geometry of a meander dipole and a half-wave dipole ($s<l$ and $s_{tot}>l$)*

The main outcome of this section is that the resonant frequency declines as some features increase such as the meander height h, the number of folds m, the meander width w and the conducting line length s. Moreover, the loaded position of the meanders does not affect the resonant frequency but strongly modifies the gain performance.

2.1.3.1. *Analytical analysis of the structure*

The analysis of the structure can be made as follows [END 00]. The influence of the meander part of the antenna is similar to a load, and the meander line sections are considered as shorted-terminated transmission lines. Figure 2.12 shows a dipole antenna with two meanders. Each meander is treated as a twin line with a short-circuited termination. In addition, the bold line and the dashed line are considered as a straight conducting line with length s and diameter b.

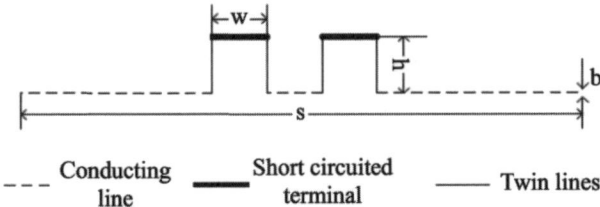

--- Conducting line	Short circuited terminal	Twin lines

Figure 2.12. *Meander line sections seen as shorted-terminated transmission lines*

The characteristic impedance of twin lines can be expressed in the following form:

$$Z_0 = \frac{\eta}{\pi} \log \frac{2w}{b}$$ [2.21]

where η is the wave impedance in free space, w is the distance between twin lines and b is the diameter of the conducting line. Z_{in} is the input impedance of twin lines, given by the following equation:

$$Z_{in} = Z_0 \frac{Z_L + jZ_0 \tan \beta h}{Z_0 + jZ_L \tan \beta h}$$ [2.22]

where β is equal to $2\pi/\lambda$ and h is the height of the twin lines. Assuming that all twin lines are terminated in a short circuit, the load impedance of the twin lines is zero ($Z_L=0$) and [2.22] becomes:

$$Z_{in} = jZ_0 \tan \beta h$$ [2.23]

where tanβh can be expanded into three orders on condition that $\beta h \ll 1$.

$$\tan \beta h \approx \beta h + \frac{1}{3}(\beta h)^3$$ [2.24]

Then a new expression of input impedance is obtained:

$$Z_{in} = j\omega L = jZ_0\left(\beta h + \frac{1}{3}(\beta h)^3\right)$$ [2.25]

If we insert [2.21] into [2.25], the reactance formed by each twin line can be shown to be:

$$L = \frac{\mu_0 h}{\pi}\ln\frac{2w}{b}\left(1 + \frac{1}{3}(\beta h)^2\right)$$ [2.26]

where μ_0 is the vacuum permeability. On the assumption that the number of meanders is m, the total reactance obtained by the twin lines should be $L_p = m \times L$. The straight conducting line, whose length is s, also results in a self-inductance. It is given by the following equation [END 00]:

$$L_s = \frac{\mu_0 s}{2\pi}\left(\ln\frac{4s}{b} - 1\right)$$ [2.27]

The total inductive reactance of the MDA is finally given by:

$$L_T = L_s + m \times L$$ [2.28]

The self-inductance of a half-wavelength dipole antenna can also be derived from [2.27]:

$$L_H = \frac{\mu_0 \lambda}{4\pi}\left(\ln\frac{2\lambda}{b} - 1\right)$$ [2.29]

Following [END 00], we suppose that the inductive reactances of the MDA and the half-wave dipole antenna are the same when they resonate at the same frequency. Thus, $L_H = L_T$ at $f_0 = c/\lambda_0$ leading to:

$$\frac{\mu_0 s}{2\pi}\left(\ln\frac{4s}{b} - 1\right) + m\frac{\mu_0 h}{\pi}\ln\frac{2w}{b}\left(1 + \frac{1}{3}(\lambda_0 h)^2\right) = \frac{\mu_0 \lambda_0}{4\pi}\left(\ln\frac{2\lambda_0}{b} - 1\right)$$ [2.30]

[2.30] states the relationship between the resonant frequency f_0 of the MDA and its physical dimensions.

2.1.3.2. *Influence of the meander characteristics on the resonant frequency*

MDAs loaded with different numbers of meanders are selected as shown in Figure 2.13, where the number of meanders $m = 2$, 8 or 14, the length of MDA s = 129 mm, the wire diameter b = 1 mm and the gap between dipole arms g = 3 mm [HU 09].

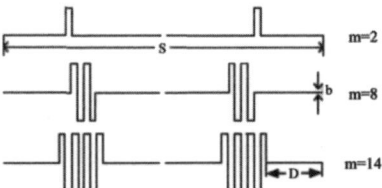

Figure 2.13. *Topology of the three MDAs under study*

In order to establish the influence brought by each parameter on the resonant frequency, the following methodology is used. Two of the three values (*m*, *h*, *w*) are fixed while the unfixed one is changed over a range. To examine the validity of the analytical model, the simulation software Ansoft high frequency structural simulator (HFSS) is used for comparison.

The number of meanders *m* is changed from 2 to 14 (*w* = 6 mm, *h* = 10 mm). The resonant frequency calculated by [2.30] and HFSS is derived, as shown in Figure 2.14. Similarly, the variation of resonant frequency caused by the meander height and width is shown in Figures 2.15 and 2.16 for *m* = 2, 8 and 14. It is observed that the original size of the half-wave dipole can be reduced by nearly 50% with 14 meanders.

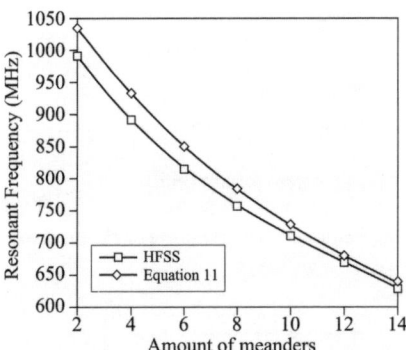

Figure 2.14. *Resonant frequency as a function of the number of meanders*
(w = 6 mm, h = 10 mm)

The influence of the positions of the meanders can be studied by moving them toward the end of the dipole (distance *D* in Figure 2.13) with unchanged meander dimensions (*w* = 6 mm, *h* = 10 mm). The resonant properties analyzed by [2.30] and HFSS are illustrated in Figure 2.17. We note that *D* does not have a strong effect on the resonant frequency. Overall, the resonant characteristics of the MDA predicted by [2.30] are in general agreement with simulation results and are getting closer with the increase of the number of meanders.

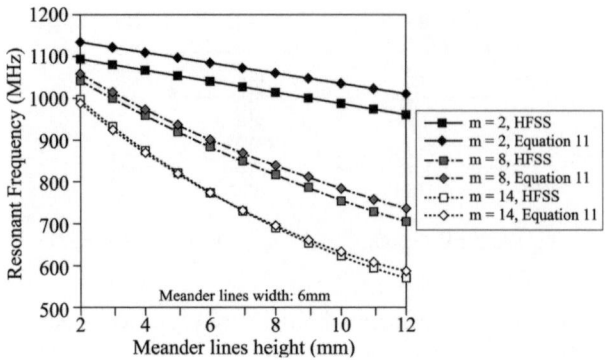

Figure 2.15. *Resonant frequency as a function of the meander height (w = 6 mm)*

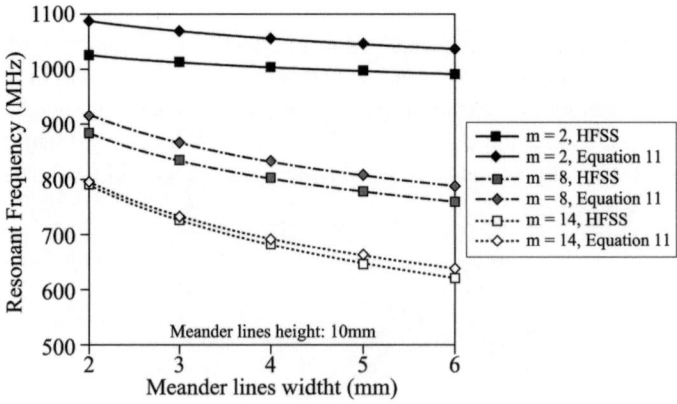

Figure 2.16. *Resonant frequency as a function of the meander width (h = 10 mm)*

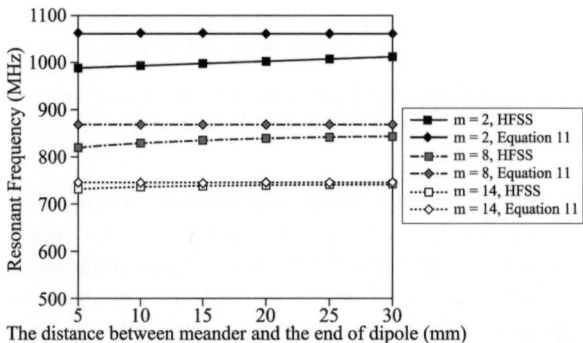

Figure 2.17. *Resonant frequency as a function of the position D of the meanders*

2.1.3.3. *Discussion on the MDA reactance*

The inductance of twin wires is lower than the inductance of a wire of the same length. For instance, assuming a diameter $b = 1$ mm, a straight wire of length $l = 100$ mm yields $L = 100$ nH with [2.27], whereas we find 60 nH for twin wires of height $h = 47.5$ mm and width $w = 5$ mm with [2.26]. Therefore, the inductance of a straight dipole will be always higher than that of an MDA if the length of the straight dipole and the total "physical" length of the MDA wire are identical. A densely packed meandered antenna has significantly less inductance per unit length than a straight dipole.

However, the total capacitance of an MDA is also reduced compared to a straight dipole. This can be deduced from the rough approximation [2.2] of the capacitance for short dipoles. Since the resonant frequency of an antenna is equal to $1/2\pi\sqrt{LC}$ where L and C are the inductance and capacitance of the antenna, respectively, we conclude that a longer "physical" length s_{tot} is required for an MDA that would have been the case using a straight dipole to obtain the same 900 MHz operation.

2.1.3.4. *Influence of the meander characteristics on the gain and the radiation resistance*

A standard half-wave dipole antenna is modeled in HFSS. Its length $l = 129$ mm is the same as that of the MDA discussed in the previous section. A comparison is made between the gain of the half-wave dipole and that of the MDA with a fixed "mechanical" length $s = 129$ mm. The gains are all derived at the half-wave dipole resonant frequency 1.095 GHz and the MDA resonant frequencies. The comparison is shown in Tables 2.1 and 2.2. It is demonstrated that the MDA gain decreases as the number of meanders, their height and width increase.

h(mm)/m	2	6	10
2	3.00	2.90	2.67
8	2.90	2.22	1.63
14	2.89	1.70	0.39

Table 2.1. *MDA gain for different numbers of meanders and meander heights (w=6 mm)*

W(mm)/m	2	3	4	5	6
2	2.86	2.81	2.77	2.72	2.67
8	2.35	2.01	1.80	1.75	1.63
14	1.64	1.20	0.80	0.62	0.39

Table 2.2. *MDA gain for different numbers of meanders and meander widths (h=10 mm)*

The gain drop with MDAs can be explained as follows. The currents on the two parallel vertical wires which compose the twin lines are flowing in opposite directions. Consequently, these currents do not radiate but contribute to the magnetic near-field and the conducting losses. In other words, a meander increases not only the antenna inductance but also its loss resistance R_{loss}. On the other hand, the uncompensated currents in the horizontal segments of the meanders contribute to the radiation. As the total lengths of horizontal currents in the MDA and the half-wave dipole are both equal to 129 mm, it can be assumed that both antennas have the same radiation resistance R_{rad}. Therefore, the antenna efficiency given by $\eta = R_{rad}/(R_{rad}+R_{loss})$ is reduced with the MDA and meanders lower the antenna efficiency and gain. In conclusion, increasing the number of meanders does not improve the radiation ability but actually results in more losses, i.e. compromise should be made between the size reduction and the gain.

Another important point is that the position of the meanders along the MDAs has some impact on the gain. As the current along a resonant dipole is minimum at its ends and maximum at its center, it is preferable to concentrate the meanders at the ends rather than in the center to reduce the losses, whereas the total inductance is independent of the meanders' position.

Conversely, if the MDA is squeezed to keep the same resonant frequency as the half-wave dipole then the radiation resistance $R_{rad,\,meander}$ of the MDA can be deduced from the radiation resistance of the half-wave dipole $R_{rad,\,half-wave}$ by [DOB 12]:

$$R_{rad,meander} \approx \left(\frac{2s}{\lambda}\right)^2 R_{rad,half-wave} \qquad [2.31]$$

2.1.4. Influence of dielectric and metallic materials – losses and detuning

This section focuses on the influence of the materials used in the fabrication of antenna tags, i.e. the conducting metal-line and the dielectric substrate. The losses due to these parameters and the impact on the antenna efficiency are first studied. Then, the detuning effect on the resonant frequency and the input impedance will be analyzed.

2.1.4.1. General formulation of the dielectric and conducting losses

It is reminded that the power dissipation in dielectric heating in a dielectric volume (V) is given by:

$$P_d = \frac{1}{2}\omega\varepsilon_0\varepsilon_r \tan\delta \iiint\limits_V |\vec{E}|^2 dv \qquad [2.32]$$

where ε_r and tanδ are the relative permittivity and the loss tangent of the dielectric, respectively. As dielectric losses increase in proportion to tanδ, we conclude that the larger the value of the loss tangent, the more energy is converted from electric field into heat. This means that if there is a lossy dielectric material near a tag, the tag will lose some received power from the reader in the dielectric and we would expect to see a reduction in the tag antenna performance. Materials having polar molecules, such as water, especially tend to have high dielectric losses. However, the challenge in having good tag performance in the presence of targets containing water is not only due to the high dielectric losses, but also due to the high relative permittivity value of water, which is approximately 81.

The power dissipation due to ohmic conduction over a metallic surface (S) is given by:

$$P_j = R_s \iint_S \left| \vec{H}_t \right|^2 dS \qquad [2.33]$$

where $\vec{H}_t$ is the component of the magnetic field tangent to the surface and R_s is the surface resistance given by:

$$R_s = \sqrt{\frac{\pi \mu_0 f}{\sigma}} \qquad [2.34]$$

We conclude that conducting losses increase in proportion to $1/\sqrt{\sigma}$. In a good conductor at high frequencies, such as UHF, current density is packed into the regions near the conductor surface. The depth below the surface of a conductor at which the amplitude of an incident electric field has decreased by a factor 1/e is called the skin depth $\delta = \sqrt{2/\sigma\omega\mu}$. The δ-value is roughly 2 μm for copper at 900 MHz. When the conductor thickness is equal or less than the skin depth, the superficial resistance becomes inversely proportional to the conductor thickness. Therefore, a printed RFID antenna whose thickness is smaller than the skin depth will have a low efficiency.

2.1.4.2. Analysis of the dielectric losses

The efficiency analysis is based on the work presented in [CHO 07] for a simple meander-type tag structure (Figure 2.1) using a T-matching network to match a commercial tag chip located at the centre of the antenna.

Three different substrate materials made of polyethylene terephthalate (PET) (εr: 3.9, tanδ: 0.03, thickness: 50 μm), Duroid (εr: 2.2, tanδ: 0.0009, thickness: 127 μm) and FR-4 (εr: 4.25, tanδ: 0.02, thickness: 1.6 mm) are used to examine the effect of the substrate material and thickness. The read-ranges for a given tag antenna size are shown in Figure 2.19. The read-range rapidly decreases as the antenna size is reduced since the radiation efficiency of the antenna drops. It is also clear that the use of high loss substrates such as FR-4 results in low antenna efficiencies and shorter read-ranges. By using thin substrates such as PET, the read-range of the tag can be greatly increased nearly to the value of the low-loss and high-cost substrate material, such as Duroid, despite the fact that the loss tangent of PET is about the same as that of FR-4. Figure 2.20 shows the radiation efficiency as a function of the substrate thickness for a given antenna size ($kr = 0.6$). As expected, the efficiency decreases as the thickness of substrate increases. For thin substrates (<0.4 mm), the loss tangent influence is reduced with a radiation efficiency over 80% for the three types of materials.

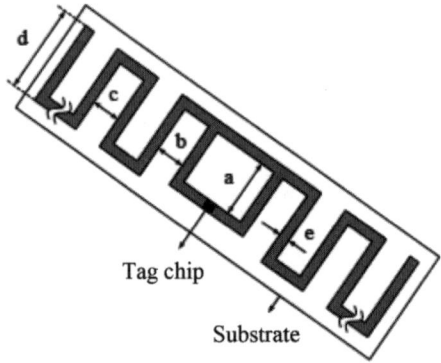

Figure 2.18. *Geometry of the meander tag antenna used in the study of dielectric losses*

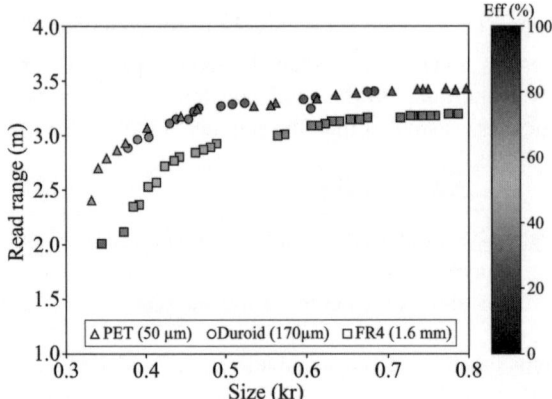

Figure 2.19. *Read-range versus antenna size and substrate materials.*
Δ: PET (50 μm); O: Duroid (170 μm); □: FR-4 (1.6 mm)

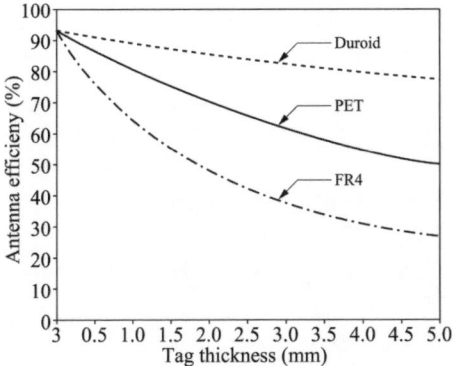

Figure 2.20. *Antenna efficiency versus substrate thickness and substrate materials.'* ———
Duroid; — – — – *PET;* — — *FR4*

Finally, the antenna efficiency in Figure 2.21 is plotted when the tag is attached on different dielectric materials (all 2 mm thick) showing different permittivities and losses. Clearly, the efficiency decreases as the permittivity and loss tangent of the tagged objects are increased. These results show that the electrical properties of the tagged objects must be taken into account when designing the tag antennas.

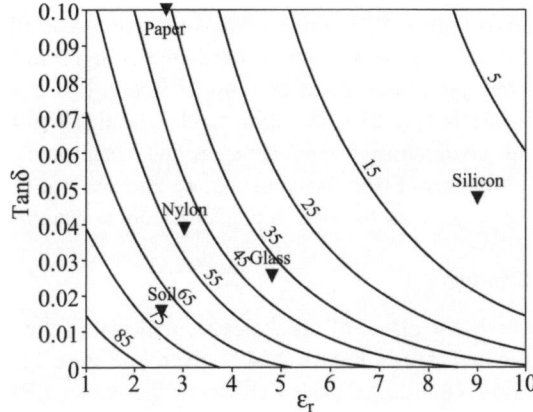

Figure 2.21. *Antennae efficiency versus dielectric properties of the tagged object*

2.1.4.3. *Analysis of the metallic losses*

As the conductivity values are $\sigma = 5.8 \times 10^7$ S/m for copper and $\sigma = 1.6 \times 10^6$ S/m for silver ink, respectively, we expect reduced performances for silver ink printing [NIK 05]. The performance comparison between silver ink printing and copper etching is shown in Figure 2.22. The degradation of read range with the silver ink becomes more significant for small antenna sizes ($kr < 0.5$).

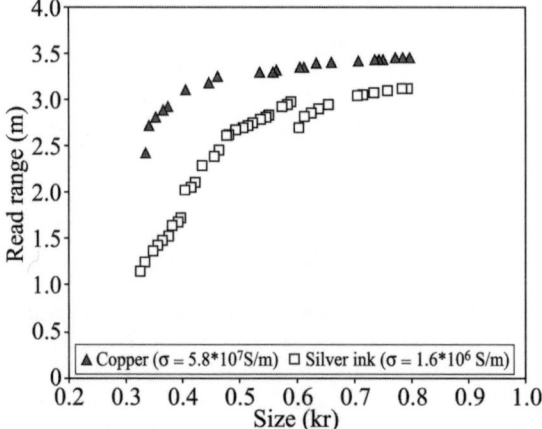

Figure 2.22. *Read range versus antenna size and metal line conductivity. Δ: copper ($\sigma = 5.8 \times 10^7 S/m$); □: silver ink ($\sigma = 1.6 \times 10^6 S/m$)*

Next, we examine the performance change due to a variation of the line thickness, as shown in Figure 2.23. The antennas are printed on PET substrate with copper metal-line whose thickness is varied between 0.1 and 5 µm. The read ranges decrease rapidly when the thickness of the copper line is less than 0.7 µm. As the thickness of metal-line is less than the skin depth (about 0.7 µm at 900 MHz for copper), significant conducting losses arise in the metal-line and the antenna efficiency is greatly reduced. From these results, we can see that the thickness of the metal-line should be larger than the skin depth for adequate readability.

2.1.4.4. *Antenna detuning*

In order to illustrate the effect of the dielectric material of the tagged object, the read range of the tag depicted in Figure 2.24 is given in Figure 2.25 as a function of frequency for an effective radiated power (ERP) of the reader ERP=2W [RAO 05]. Three configurations are considered: tag in free-pace, tag on Box 1, an empty cardboard box, and tag on Box 2, a cardboard box with plastic material $\varepsilon r = 2.87$ inside. The read-range responses are shifted downward when the tag is put on either box because of the dielectric loading. Then, the tag is tuned by reducing the loading bar and the meander, as shown in Figure 2.24. The tuning differs for each box. The objective is to shift the read-range peak upward in the 868 MHz band. The presented tag design meets the desired range requirements (>2 m) for both boxes in 868 and 915 MHz frequency bands.

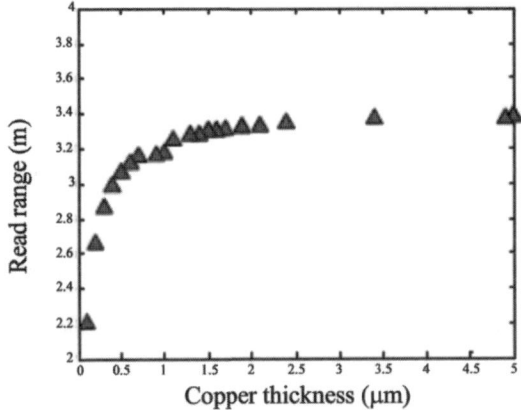

Figure 2.23. *Read range versus metal line thickness*

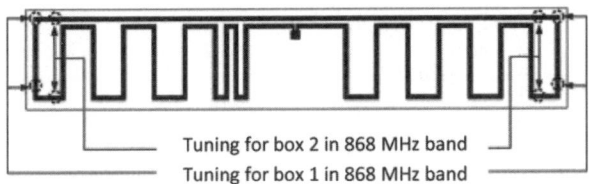

Figure 2.24. *RFID tag using a loaded meander antenna. Tuning of the meanders and the bar for various box contents in 868 MHz band [RAO 05]*

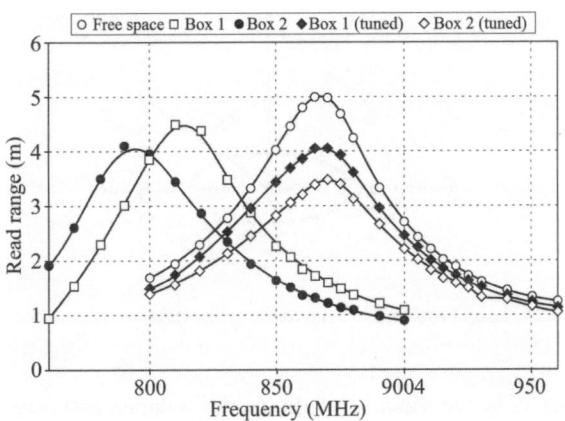

Figure 2.25. *Variations of RFID tag range versus frequency for different tagged objects (ERP = 2 W). Box 1: empty cardboard box. Box 2: cardboard box with plastic material ($\varepsilon r = 2.87$) inside*

2.1.5. Near-field/far-field behavior of UHF RFID tags

In the previous sections, we have observed that a tag antenna can eventually be seen as the connection of a dipole-like antenna with a loop conductor. This loop is the result of the symmetrical association of two L-matching circuits. Concurrently, any small loop can be seen as a magnetic field sensor, whose sensitivity increases in proportion to the loop surface as shown below.

Let us consider the geometrical arrangement of the loop antenna of radius a, shown in Figure 2.26. A constant current I flows around the loop, which lies in the xy-plane and is centered at the origin. The near-field components are given by:

$$
\left.
\begin{aligned}
H_r &\approx \frac{a^2 I e^{-jkr}}{2r^3} \cos\theta \\
H_\theta &\approx \frac{a^2 I e^{-jkr}}{4r^3} \sin\theta \\
H_\varphi &= E_r = E_\theta = 0 \\
E_\phi &\approx -j\frac{a^2 k\eta_0 I e^{-jkr}}{4r^2} \sin\theta
\end{aligned}
\right\} \quad kr \ll 1
$$

[2.35]

where $k = 2\pi/\lambda$ is the propagation constant in free space, η_0 is the wave impedance of free-space, r is the distance between the loop center and the observation point, and θ is the angle between the z-axis and the direction of the observation point. H_r and H_θ vary as $1/r^3$ while E_ϕ varies as $1/r^2$. Therefore, as r is small, the H-field dominates the E-field in the near-field and the loop acts as a magnetic sensor.

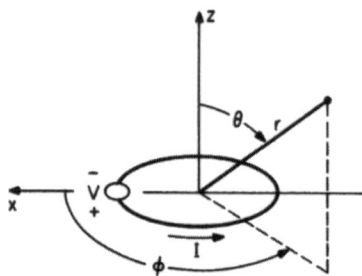

Figure 2.26. *Loop antenna and accompanying spherical coordinate system*

When the tag is in the vicinity of the reader antenna (say, up to 10 cm), it is actually located in the near-field of the reader antenna. The IC powering is mainly due to the magnetic flux across the loop while the dipole impact is weak. For larger distances, when the tag is in the far-field of the reader antenna, the dipole becomes a far-field sensor boosting the radiation resistance of the structure. This dual-behavior is stressed in Figure 2.27 where the loop is depicted as the near-field element and the dipole as the far-field element.

The same dual-behavior is used in a module combining a small loop and an RFID chip, the AK tag developed by the French company Tagsys (Figure 2.28). In a standard use, the AK tag is used for short-range communications (<50 cm), but its read range can be extended up to 10 m when the module is electromagnetically coupled to a resonant piece of metal. As the loop is used as a primary source and not as a dedicated matching element, bandwidth performance might be reduced, but the concept is so flexible and versatile that numerous industrial applications are possible.

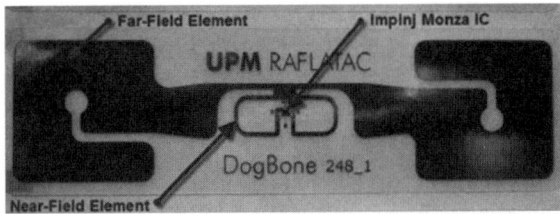

Figure 2.27. *Dual-behavior of a tag antenna: far-field and near-field elements*

Figure 2.28. *AK tag (Tagsys) using the UHF Gen 2 chip Monza 5. Frequency band 860–960 MHz; inlay dimensions: 12 mm × 10 mm*

2.2. Matching between the antenna impedance and the microchip impedance

2.2.1. *Matching conditions*

Let us assumes that the IC impedance takes the form of the series combination of a resistance R_{ic} and a reactance X_{ic} and that the antenna is modeled by a linear source voltage V_a, the source impedance being $Z_a = R_a + jX_a$ (see Figure 2.29). The power P_{ic} delivered to the load R_{ic} by the source is given by:

$$P_{ic} = \frac{1}{2} V_{ic} I_a^* = \frac{V_a^2}{2} \frac{R_{ic}}{\left| Z_a + Z_{ic} \right|^2} \qquad [2.36]$$

where I_a is the current flowing along the circuit. The maximum power transfer between the antenna and the load occurs if the matching conditions are fulfilled, i.e. when the source and load impedances are complex conjugates, $R_{ic} = R_a$ and $X_{ic} = -X_a$. Under matching conditions, the power P_{max} delivered to the tag IC is given by:

$$P_{max} = \frac{R_{ic} I_a^2}{2} = \frac{V_a^2}{8 R_a} \qquad [2.37]$$

The power transfer coefficient (PTC) τ is then defined as:

$$\tau = \frac{P_{ic}}{P_{max}} = \frac{V_a^2}{2} \frac{R_{ic}}{\left| Z_a + Z_{ic} \right|^2} \frac{8 R_a}{V_a^2} = \frac{4 R_{ic} R_a}{\left| Z_a + Z_{ic} \right|^2} \qquad [2.38]$$

The power delivered to the tag and the tag read-range are maximized for $\tau = 1$. The magnitude $|\Gamma|$ of the voltage reflection coefficient can be deduced from τ using

$$|\Gamma| = \sqrt{1 - \tau} = \left| \frac{Z_a - Z_{ic}^*}{Z_a + Z_{ic}} \right| = \frac{\sqrt{(R_a - R_{ic})^2 + (X_a + X_{ic})^2}}{\sqrt{(R_a + R_{ic})^2 + (X_a + X_{ic})^2}} \qquad [2.39]$$

while the power reflection coefficient is $|\Gamma|^2$ and the return loss is given by $RL(dB) = -20 \log |\Gamma|$.

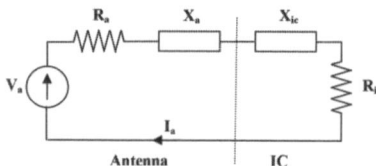

Figure 2.29. *Circuit model of tag antenna and IC*

The tag IC is usually represented as the parallel combination of some capacitance C_{cp} and some resistance R_{cp} that are related to the series capacitor C_{ic} and resistance R_{ic} through the following relationships:

$$R_{ic} = \frac{R_{cp}}{1 + R_{cp}^2 C_{cp}^2 \omega^2} \approx \frac{R_{cp}}{Q_{cp}^2} \qquad [2.40]$$

$$C_{ic} = \frac{1 + R_{cp}^2 C_{cp}^2 \omega^2}{R_{cp}^2 C_{cp} \omega^2} \approx C_{cp} \qquad\qquad [2.41]$$

Typical values provided by the main manufacturers (Alien Technologies, Impinj, NXP semiconductor, etc.) are around 1 kΩ for R_{cp} and 1 pF for C_{cp}. Therefore, the rough order of magnitude estimate is 7 for $Q_{cp} = R_{cp}C_{cp}\omega$ and 30 Ω for R_{ic} in the UHF band.

2.2.2. L-matching basics

The great majority of commercial UHF RFID tags are based on dipole antennas using a modification of a T-match as a matching circuit. Most of the time, a T-match reduces to an L-match (Figure 2.30) as one of the series inductors can readily be included in the antenna reactance.

The series resonant frequency f_a of the dipole is normally located above the working frequency f_0 for miniaturization purposes. Also, $f_a > f_0$ allows us to reduce the influence of surrounding materials showing moderate permittivities. As a result, the dipole impedance at f_0 is capacitive with typical resistances of a few dozen ohms and reactances ranging from roughly 100 to 200 Ω.

It is demonstrated in Figure 2.30 that an L-matching circuit based on two inductances L_e and L_h is sufficient to match the impedance Z_{ic} of any UHF RFID IC with most dipole antennas found in RFID tags. Because $\text{Re}(Z_{ic})$ is small, the antenna impedance is always included inside the $\text{Re}(Z_{ic}) = $ constant resistance circle. As a result, a proper combination of series and parallel inductances L_e and L_h create the complex conjugate $Z_{ic}*$ of the chip impedance. Typical inductances values range from few nH to few tens of nH and can be readily realized in microstrips.

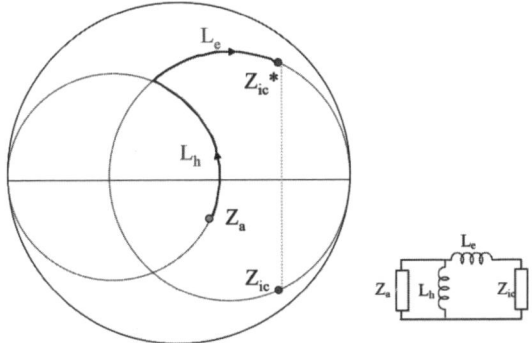

Figure 2.30. *Antenna matching to the UHF RFID IC with an L-match*

2.2.3. *Equivalent electrical circuits*

Figure 2.31 shows a typical matching circuit for a commercial RFID tag with superimposed currents. In [DEA 09], the balanced circuit shown in Figure 2.31 can be fairly closely mapped to the unbalanced circuit shown in Figure 2.32, but mutual inductive couplings between L_e and L_h and possibly L_a are neglected. Fortunately, any inductive coupling will serve to transform the antenna impedance, and will not, for example, disturb the antenna Q or resonant frequency. Thus, we can conclude that Figure 2.32 is qualitatively correct, although an exact translation between circuit element values and physical geometry is lost.

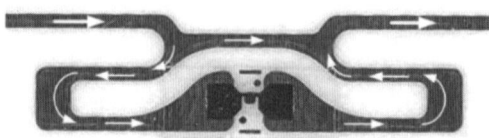

Figure 2.31. *Commercial tag matching circuit with superimposed currents [DEA 10]*

A series RLC circuit is used to model the antenna near resonance. Its resonant pulsation ω_a and quality factor Q_a are given by the well-known relationships:

$$\omega_a = \frac{1}{\sqrt{L_a C_a}} \qquad\qquad [2.42]$$

$$Q_a = \frac{\omega_a L_a}{R_a} = \frac{1}{\omega_a R_a C_a} \qquad\qquad [2.43]$$

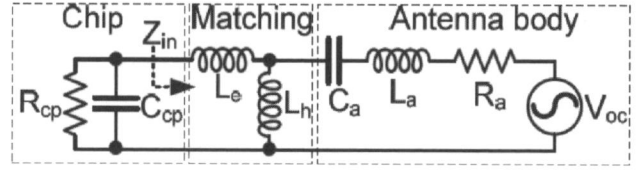

Figure 2.32. *Equivalent circuit of the double-tuned unbalanced tag*

The circuit is then transformed into a more useful form as depicted in Figure 2.33 using the L-match transform identity in which the shunt-series inductors (L_e and L_h) are replaced by the series-shunt inductors (L_{se} and L_n). The result of this transformation is a circuit that is in the canonical form of a two-stage bandpass filter with a series resonant circuit (R_{a2}, $L_{a2} + L_{se}$ and C_{a2}) and a parallel resonant circuit (R_{cp}, C_{cp} and L_n):

where

$$\alpha = L_h / (L_e + L_h)$$ [2.44]

$$R_{a2} = R_a / \alpha^2$$ [2.45]

$$L_{a2} = L_a / \alpha^2$$ [2.46]

$$C_{a2} = \alpha^2 C_a$$ [2.47]

$$Q_{a2} = L_{a2}\omega_a / R_{a2} = 1/(R_{a2}C_{a2}\omega_a) = Q_a$$ [2.48]

$$L_n = L_e + L_h$$ [2.49]

$$L_{se} = L_e / \alpha$$ [2.50]

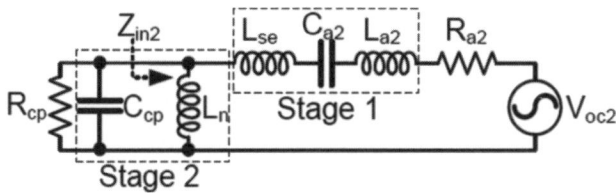

Figure 2.33. *Equivalent circuit of the double-tuned unbalanced tag after L-match transformation. Circuit in the form of a two-stage bandpass filter*

From Figure 2.33, we observe that the parallel resonance is independent of the radiating part. Therefore, the parallel resonance will not be directly affected by the dielectric environment of the tag. On the other hand, the series resonance will shift toward lower frequencies as the antenna electrical length increases with the dielectric permittivity.

2.2.4. *Double-tuned matching*

The practical implication of the circuit model described in section 2.2.3 is that a desired bandpass response can be synthesized with a proper calculation of the antenna impedance and the matching circuit [DEA 09]. We assume that the design parameters given are:

– the antenna form factor that largely determines Q_a and R_a;

– the IC input impedance.

By tailoring the classic double-tuning theory to fit the model of Figure 2.33, the first condition of the optimum double-tune matching is to select the resonant frequency of the series and parallel tank circuits to be the same:

$$\frac{1}{\sqrt{L_n C_{cp}}} = \frac{1}{\sqrt{(L_{se} + L_{a2})C_{a2}}} = \omega_0 \tag{2.51}$$

where ω_0 is equal to the geometric mean of the band, i.e. $\omega_0 = \sqrt{\omega_{min} \omega_{max}}$. It can be seen that L_n is entirely fixed by the chip reactance once ω_0 is given.

α is the remaining tuning parameter. Any change in α will change L_{se} and, thus, the series resonance. If we wish to change α, we will also need to modify ω_a, the resonant frequency of the antenna, so that the resonant frequency of the series (R_{a2}, $L_{a2} + L_{se}$ and C_{a2}) circuit remains ω_0. By combining [2.42] and [2.51], we obtain the following quadratic in ω_a:

$$\omega_a^2 - \omega_a \frac{\omega_0^2 L_{se}}{R_{a2} Q_a} - \omega_0^2 = 0 \tag{2.52}$$

which is solved in the normal way. [2.52] can be rewritten in terms of L_a using [2.43]:

$$L_{a2}^2 + L_{a2} L_{se} - \left(\frac{R_{a2} Q_a}{\omega_0}\right)^2 = 0 \tag{2.53}$$

Finally, by applying the quadratic equation and considering only the positive inductances, we obtain the synthesized antenna inductance:

$$L_{a2} = \frac{\sqrt{L_{se}^2 + \left(\frac{2 R_{a2} Q_a}{\omega_0}\right)^2} - L_{se}}{2} \tag{2.54}$$

or equivalently:

$$L_a = \frac{\sqrt{(\alpha L_e)^2 + \left(\frac{2 R_a Q_a}{\omega_0}\right)^2} - \alpha L_e}{2} \tag{2.55}$$

Finally, inductances are constructed from metallic flat or ribbon wire (rectangular cross section), using the following estimation:

$$L(nH) = 0.002\,l\left(\ln\frac{l}{w+t}+1.193+0.2235\frac{w+t}{l}\right)$$ [2.56]

where w and l are, respectively, the width and the length of the trace and t is the metallization thickness (all dimensions in cm). Once the previous values are fixed, it is possible to determine the reflection coefficient at $\omega = \omega_0$ when both series and parallel tank circuits resonate:

$$|\Gamma| = \frac{R_{cp}-R_{a2}}{R_{cp}+R_{a2}} = \frac{R_{cp}-R_a/\alpha^2}{R_{cp}+R_a/\alpha^2}$$ [2.57]

or in terms of PTC:

$$\frac{R_{cp}}{R_{a2}} = \frac{2}{\left(1-\sqrt{1-\tau}\right)}-1.$$ [2.58]

In [2.57] and [2.58], $|\Gamma|$ and τ are the worst reflection coefficient and the worst PTC, respectively, allowed through the bandwidth and observed at $\omega = \omega_0$.

The optimum double-tuned match provides the maximum fractional bandwidth out of double-tuning [LOP 07]:

$$BW_{opt} = \frac{2\sqrt{|\Gamma|}}{Q_a\left(1-|\Gamma|\right)\sqrt{\alpha\dfrac{L_e}{L_a}+1}}$$ [2.59]

2.2.5. Synthesis of a double-tuned tag and a naïve tag

Let us assume an antenna with $Q_a = 15$ which is approximately that of a meandering dipole with a length of 92 mm and a width of 8 mm. This size is popular in the high-volume commercial market, and thus conclusions drawn from this example are of commercial relevance. We let a typical $R_a = 30\ \Omega$ and we assume the use of an IC with $R_{cp} = 1500\ \Omega$ and $C_{cp} = 1.2$ pF for $Q_{cp} = 10.3$. The goal is to maintain a 10 dB return loss, i.e. a power transfer efficiency of 90% between 860 and 960 MHz (worldwide operation in the Gen2 standard definition).

RL = 10 dB means that $|\Gamma|$=0.316 at f_0 = $\omega_0/2\pi$ = 908.6 MHz, the geometric mean of the UHF band. From [2.57], we obtain α = 0.196. Now $\omega_0 = 1/\sqrt{C_{cp}L_n}$ implies that $L_e + L_h = 1/C_{cp}\omega_0^2$ = 25.57 nH, and using [2.44] we find that L_h = 5.01 nH and L_e = 20.56 nH. Knowing L_{se} and using [2.52] and [2.55], it is easy to find L_a=76.83 nH and f_a = $\omega_a/2\pi$ = 931.9 MHz to achieve Q_a=15 and R_a=30 Ω. All circuit parameters including the optimized matching circuit (L_e and L_h) are summarized in Table 2.3.

The next step is to design an antenna which resonates at 931.9 MHz with Q_a = 15 and R_a = 30 Ω and fulfills the dimension constraints (92 mm × 8 mm). The wideband antenna geometry of Figure 2.34 proposed in [DEA 10] respects the conditions with a large number of meanders to achieve resonance. The simulated impedance Z_{in} of the tag circuit (antenna + L-match) was obtained from an electromagnetic simulation of the tag using a method-of-moments (MoM) code. An excellent agreement was obtained between the simulated tag impedance and the tag impedance calculated with the circuit model.

Figure 2.34. *Geometry of the wideband RFID tag*

	Q_a	R_a (Ω)	f_a (MHz)	L_e (nH)	L_h (nH)	α
Wideband	15	30	932	20.56	5.01	0.196
Naive	15	30	1,050	9.53	9.77	0.506

Table 2.3. *Circuit parameters for the wideband and naïve antennas*

The tag impedance calculated with the circuit model is plotted in Figures 2.35 and 2.36, along with the conjugate IC impedance. The imaginary parts X_{in} and X_{ic} cancel out at three frequencies, i.e. 867, 902 and 948 MHz while the real parts R_{in} and R_{ic} are equal at 879 and 935 MHz. Therefore, a perfect match does not occur at any frequency. The associated voltage reflection and power transmission coefficients are shown in Figure 2.37. The maximum transmission occurs at 876 and 942 MHz while a relative maximum of −10 dB is observed in-between at 908.6 MHz. The antenna does not achieve the 10 dB return loss over the entire 860–960 MHz band but covers 865–955 MHz, which is the practical band of worldwide operation. Both minima of the return loss correspond to frequencies where the IC and tag impedances partially match.

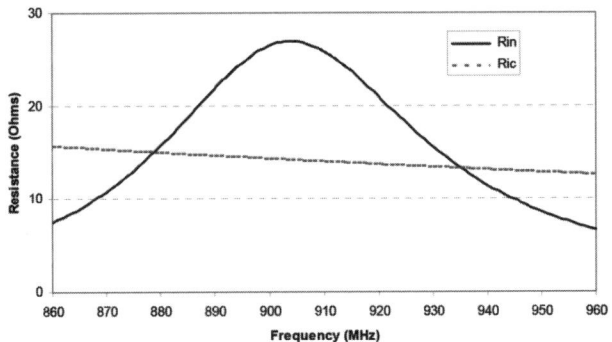

Figure 2.35. *Resistance of the wideband tag and the IC*

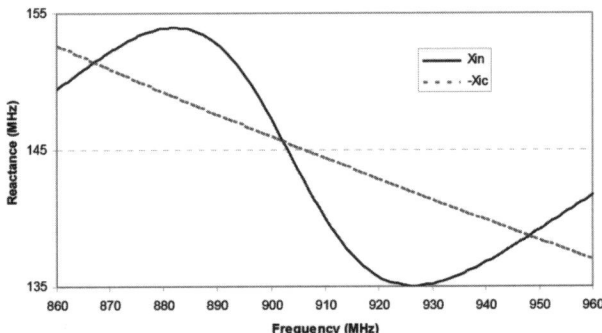

Figure 2.36. *Reactance of the wideband tag and complex conjugate of the IC reactance*

For comparison, a "naïve" short antenna resonating at a much higher frequency $f_a = 1{,}050$ MHz is designed to present a conjugate match to the IC at $f_0 = 908.6$ MHz. This antenna presents the same parameters $Q_a = 15$ and $R_a = 30\ \Omega$ as the wideband antenna. From [2.43], we extract the two remaining parameters $L_a = 68.2$ nH and $C_a = 0.337$ pF.

Using the method described in section 2.2.2, we first calculate L_h so that $\mathrm{Re}(Z_a/j\,L_h\omega_0) = \mathrm{Re}(Z_{ic})$ at f_0. We find $L_h = 9.77$ nH. Then, L_e is obtained from:

$$\mathrm{Im}\left(\frac{jZ_aL_h\omega_0}{Z_a + jL_h\omega_0}\right) + jL_e\omega_0 = \mathrm{Im}(Z_{ic}^*).$$

[2.60]

which yields L_e = 9.53 nH. The resulting input impedance Z_{in} is plotted on Figures 2.38 and 2.39 along with the conjugate IC impedance while the corresponding power reflection and power transmission coefficients are plotted on Figure 2.37. The naive design shows a perfect match at the nominal frequency 908.6 MHz but much more narrowband behavior than the wideband tag and provides only 27 MHz of 10 dB return loss. This suggests an improvement in bandwidth by a factor of about 3.3 using the double-tuning method.

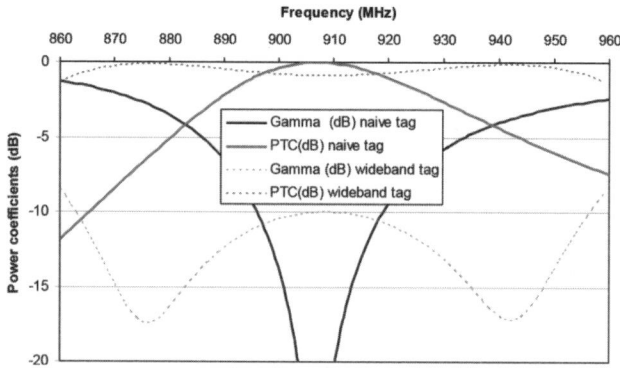

Figure 2.37. *Power transfer coefficient (PTC) and power reflection coefficient (Γ) versus frequency for the wideband tag (plain line) and the naïve tag (dotted line)*

Obviously, we can achieve much better return loss over a narrower frequency band than the wideband tag. For example, a return loss of 14 dB can be achieved between 865 and 930 MHz (covering the vast majority of worldwide operations, including North America and Europe) using the proposed approach.

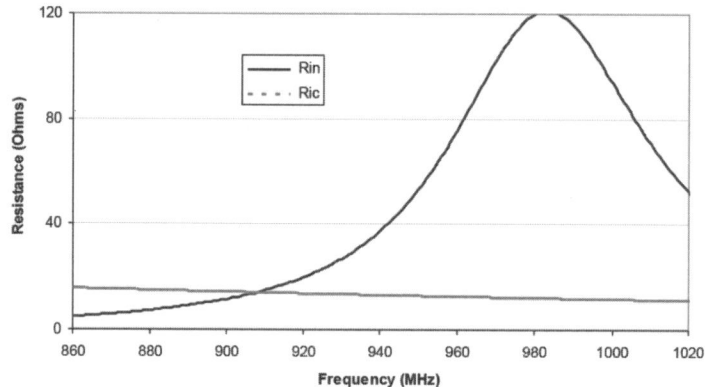

Figure 2.38. *Resistance of the naïve tag and the IC*

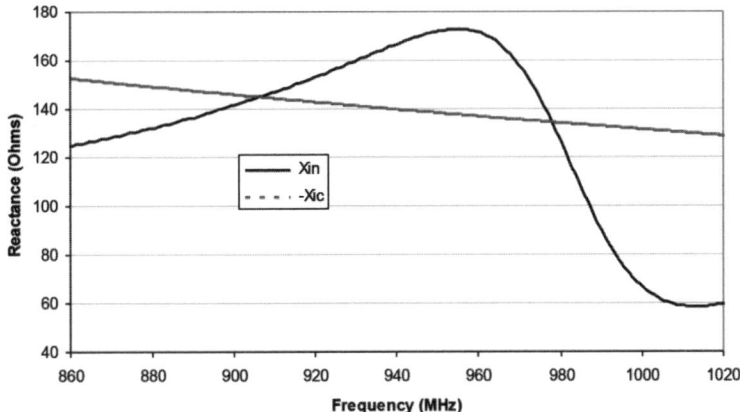

Figure 2.39. *Reactance of the naïve tag and complex conjugate of the IC reactance*

2.2.6. *Alternative implementation of the optimum double-tuned match*

The tag synthesis described in the previous section relies on the ability to design an antenna from the knowledge of the lumped-elements R_a, L_a and C_a. As a result, circuits used to model complex antenna geometries are useful abstractions but it is not easy to synthesize a specific geometry from a circuit. At least, it is beyond the capability of commercial full-wave simulators. On the other hand, the design approach proposed in [XI 11] is based on the input impedance of the whole tag antenna (i.e. the antenna body plus the matching network). According to Figure 2.32, the resistance and reactance of the tag antenna are given by:

$$R_{in} = \frac{R_a}{\dfrac{\left(R_a^2 + X_a^2\right)}{\left(\omega L_h\right)^2} + \dfrac{2X_a}{\left(\omega L_h\right)} + 1} \qquad [2.61]$$

$$X_{in} = \frac{\omega L_h R_a^2 + \omega L_h X_a \left(X_a + \omega L_h\right)}{R_a^2 + \left(X_a + \omega L_h\right)^2} + \omega L_e \qquad [2.62]$$

By taking the first-order derivate of R_{in} with respect to ω, it is found that R_{in} maximizes at the frequency given by

$$\omega_{max} = \frac{\omega_a}{1 - \dfrac{1}{2Q_a^2} + \dfrac{L_h}{L_a}} \qquad [2.63]$$

The corresponding values of R_{in} and X_{in} at $\omega = \omega_{max}$ are given by

$$R_{in-max} = R_{in}(\omega = \omega_{max}) = \frac{R_a}{\left(1 + \frac{L_a}{L_h}\right)^2 - \left(\left(1 - \frac{1}{(2Q_a^2)}\right)\frac{L_a}{L_h} + 1\right)^2} \qquad [2.64]$$

$$X_{in-max} = X_{in}(\omega = \omega_{max}) \approx \omega_{max}(L_e + L_h) \qquad [2.65]$$

Using these two expressions, the following design procedure is proposed in [XI 11]:

1) Given the antenna body and according to [2.63], L_h can be tuned to make $\omega_{max} \approx \omega_0$.

2) R_{in_max} becomes fixed with respect to [2.64].

3) L_e can be adjusted in [2.65] to make X_{in_max} approach the conjugate of the reactance of the tag chip.

4) A quick calculation of the PTC between R_{in_max} and R_{ic} reveals whether the PTC specification is satisfied at $\omega = \omega0$. If not, the antenna body should be modified. Note that the reactance of the tag antenna and that of the tag chip have cancelled out so that $\tau = 4R_{ic}R_{in-max}/(R_{in-max} + R_{ic})^2$.

2.2.7. Example of a double-tuned match tag and use in variable environments

To validate the alternative implementation described in 2.2.6, a double-tuned antenna has been built in [XI 11]. As shown in Figure 2.40, the proposed tag antenna is a loaded meander line dipole made in copper, with a total size of 93 mm × 11 mm. The matching network adopts a T-match structure, where W_1 can be adjusted to tune L_h. W_2 and H_2 can be adjusted to change L_e. The loading bar (i.e. the horizontal trace beneath the meander line) provides extra freedom to control Ra. The tag chip used in this paper is Alien Higgs-3 (R_{cp} =1.5 kΩ, C_{cp}= 0.85 pF). A parallel stray capacitance of 0.3 pF is included to account for the imperfect connection between the tag antenna and the chip strap.

A method-of-moments (MOM) full-wave simulator, IE3D, is used to assist the design. The proposed antenna is optimized for working on paper characterized by εr=2.3, loss tangent=0.1 and thickness=0.16 mm. The optimized dimensions and lumped-element model of the proposed antenna are listed in Tables 2.4 and 2.5, respectively. The lumped-element model is extracted from simulation results by curve-fitting. Results of modeling, simulation and measurement are compared in Figures 2.41 and 2.42 with respect to antenna impedance and return loss, respectively. The double-tuning effect can be clearly seen in the return loss plot. The 10-dB bandwidth is estimated to be 9.3% by implementing the data of Table 2.3 in equation [2.21], which is close to the measured result – 10.8%, which is close to the measured result – 10.8%. The difference between simulation/modeling and measurement is mainly due to the additional loss in the measurement and the uncertainty in the electromagnetic properties of the paper substrate.

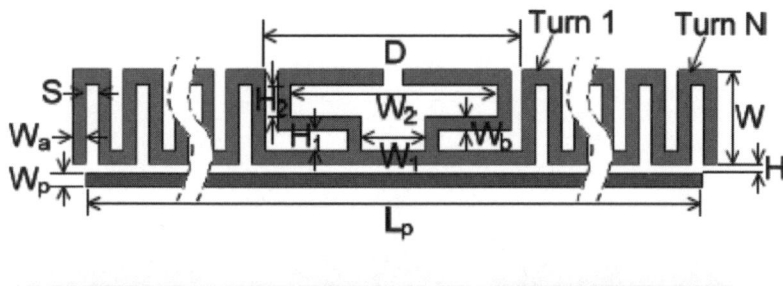

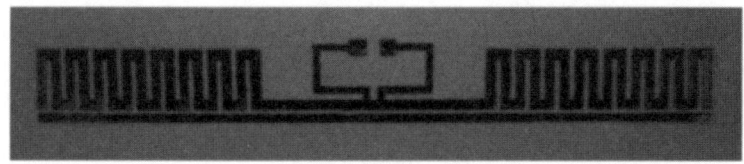

Figure 2.40. *Geometry of the proposed antenna: a) parameter definitions and b) photo of the prototype*

D	L_p	W_p	W_a	S	H	W	W_1	H_1	W_2	H_2	W_b
31	93	1	1.2	0.8	0.7	8	1.2	1	14.5	5	0.8

Table 2.4. *Dimensions of the antenna (in mm)*

R_a (Ω)	L_a (nH)	C_a (pF)	L_h (nH)	L_e (nH)	f_{a0} (MHz)	Q_{a0}	α
17.99	52.11	0.550	3.93	22.72	939.79	17.10	0.148

Table 2.5. *Lumped-element model of the antenna*

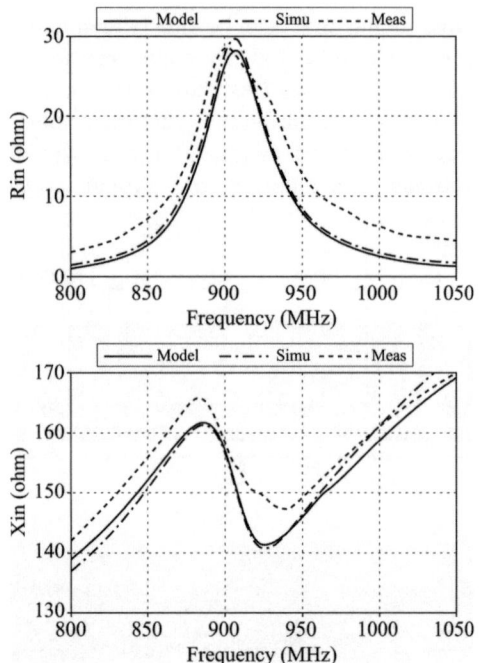

Figure 2.41. *Input impedance of the proposed antenna*

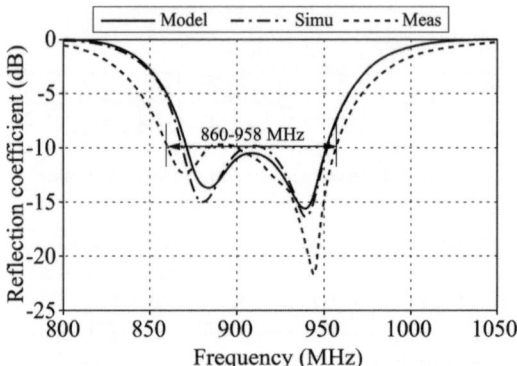

Figure 2.42. *Return loss of the proposed antenna*

The sensitivity to the dielectric loading is studied by both simulation and measurement. The simulation studies the influence of the dielectric permittivity on the bandwidth. During this simulation, loss tangent and thickness of the dielectrics are fixed as 0.1 and 0.16 mm, respectively. As shown in Figure 2.43, the acceptable matching (i.e. return loss < −6 dB) tends to disappear when the permittivity is larger than 4. Since the infinite dielectric model used in the IE3D simulator generally overestimates the dielectric loading effect, the proposed antenna should be more tolerant than what Figure 2.43 indicates.

The chip strap is mounted onto the tag antenna with conductive adhesives and the maximum reading distance is estimated with the Tagformance RFID tester from Voyantic [OCO 09]. In Figure 2.44, the measured tag responses are plotted within the world UHF RFID band (i.e. 860~960 MHz) on diverse materials. Read ranges generally stay between 8~12 m except on a book where high loss is encountered. Figure 2.44 also illustrates the double-tuning effect since the shape of the read range curves follows that of the PTC.

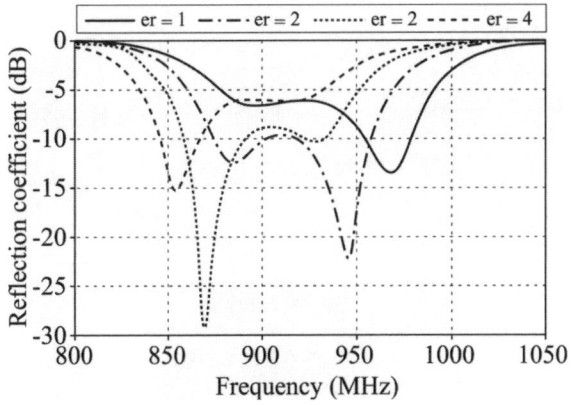

Figure 2.43. *Simulated return loss of the proposed antenna on dielectrics of different permittivities*

2.3. RFID tag antennas using an inductively coupled feed [SON 05]

This section presents a design methodology to make efficient and wideband RFID tag antennas using an inductively coupled feed. An analytical model for the inductively coupled feed is presented first. Then, the main rules to achieve wideband impedance match between the antenna and the chip are presented.

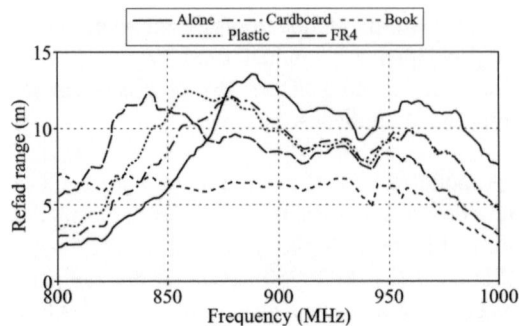

Figure 2.44. *Measured read-range on diverse materials*

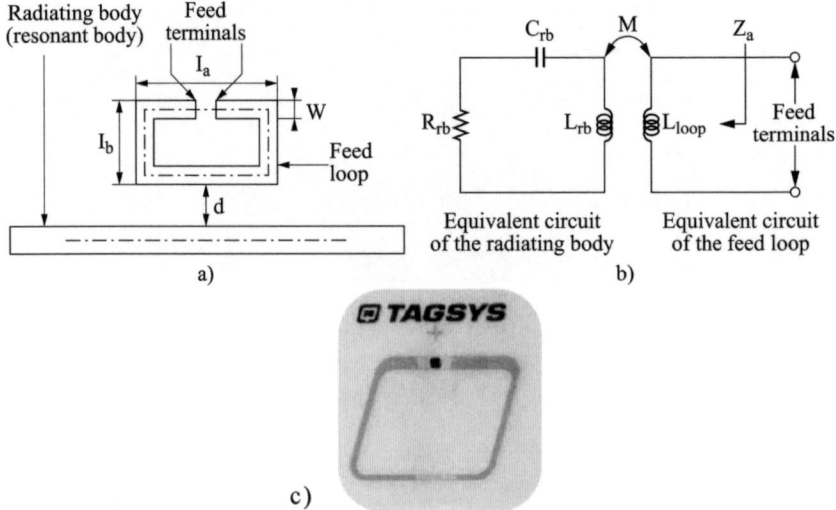

Figure 2.45. *a) Inductively coupled feed structure, b) equivalent circuit, c) example of AKA kernel (loop + IC) commercialized by Tagsys RFID*

2.3.1. *Analytical model*

The proposed feed structure is shown in Figure 2.45(a) along with dimensional notation. The antenna is composed of a small rectangular loop and a radiating (or resonant) body, which are coupled inductively. Two terminals of the loop are directly connected to the chip. The strength of the coupling is controlled by the distance between the loop and the radiating body, as well as the shape of the loop. Figure 2.45(b) also depicts the equivalent circuit of the inductively coupled feed structure. The inductive coupling is modeled by a transformer. The input impedance of the antenna Z_a is

$$Z_a = R_a + jX_a = Z_{loop} + \frac{(M\omega)^2}{Z_{rb}}$$ [2.66]

where Z_{rb} and Z_{loop} are the individual impedances of the radiating body and the feed loop, respectively. M is the mutual inductance between them, which can be roughly derived analytically on the assumption that the radiating body is infinitely long.

$$M = \frac{\mu_0}{2\pi} l_a \ln\left(1 + \frac{l_b}{d}\right)$$ [2.67]

Near the resonant frequency f_0 of the radiating body, its impedance can be expressed using the radiation resistance $R_{rb,0}$ and the quality factor Q_{rb} as a function of

$$Z_{rb} = R_{rb,0} + jR_{rb,0}Q_{rb}\left(\frac{\omega}{\omega_0} - \frac{\omega_0}{\omega}\right)$$ [2.68]

The impedance of the feed loop is as follows:

$$Z_{loop} = jL_{loop}\omega$$ [2.69]

where L_{loop} is the self-inductance of the feed loop. The resistance and reactance components of Z_a are finally given by:

$$R_a = \frac{(M\omega)^2}{R_{rb,0}} \frac{1}{1+u^2}$$ [2.70]

$$X_a = L_{loop}\omega - \frac{(M\omega)^2}{R_{rb,0}} \frac{u}{1+u^2}$$ [2.71]

where $u = Q_{rb} \cdot (\omega/\omega_0 - \omega_0/\omega)$. At $\omega = \omega_0$, the components of $Z_{a,0}$ become:

$$R_{a0} = R_a(\omega = \omega_0) = \frac{(M\omega)^2}{R_{rb,0}}$$ [2.72]

$$X_{a0} = X_a(\omega = \omega_0) = L_{loop}\omega_0$$ [2.73]

Equations [2.72] and [2.73] show that $R_{a,0}$ depends only on M, while $X_{a,0}$ is dependent only on L_{loop}. Therefore, $R_{a,0}$ and $X_{a,0}$ can be adjusted independently. This means that the proposed feed structure presents a simple and easy way to match an antenna impedance to an arbitrary chip impedance $Z_c = R_c + jX_c$. Figure 2.45(c) depicts a universal low-cost "adaptive kernel" that can act as a high-performing secondary antenna.

2.3.2. *Antenna design and results*

An example of the antenna using the proposed feed structure is shown in the inset of Figure 2.46 with the detailed dimensions. The antenna was designed for an RFID tag chip with $Z_c = 6.2-j127$ Ω. The resonant frequency f_0 of the radiating body is 915 MHz. Simulated data of $R_{rb,0} = 28.5$ Ω and $Q_{rb} = 14.7$ are obtained using CST MW Studio.

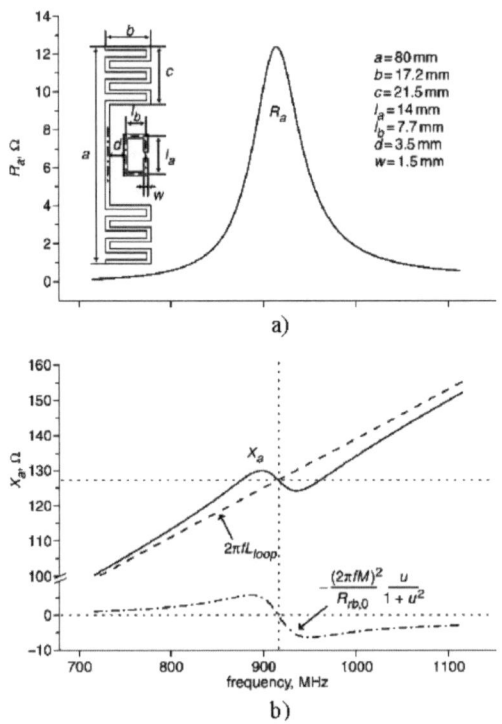

Figure 2.46. *Antenna impedance versus frequency: a) resistance component R_a and b) reactance component X_a*

The calculated values of M and L_{loop} are 3.3 and 22.1 nH, respectively. Figure 2.46(a) shows the resistance component R_a as a function of frequency, for which the maximum value is $R_{a,0} = 2R_c$ at $f = f_0$. Figure 2.46(b) shows the reactance component X_a, which equals $-X_c$ at $f = f_0$. The first term of the right side in equation [2.72] is a straight line with positive slope and the second term has negative slope in the range $f_0 \pm f_0/(2Q_{rb})$. Therefore, the first and second terms cancel each other, and it makes the inductively coupled feed structure have wideband characteristics.

VSWR versus frequency is shown in Figure 2.47 with variation of $R_{a,0}$, where $R_{a,0}$ can be adjusted by simply altering the distance between the radiating body and the feed loop without any change of $X_{a,0}$. The plot indicates that the impedance bandwidth for VSWR < 2 becomes maximum value when $R_{a,0} = 2R_c$. The antenna was fabricated and measured for validation. It was printed on a thin flexible polyethylene substrate with a thickness of 50 μm using copper traces with a thickness of 18 μm. Figure 2.47 also shows the measured VSWR of the antenna, which closely agrees with the calculated one.

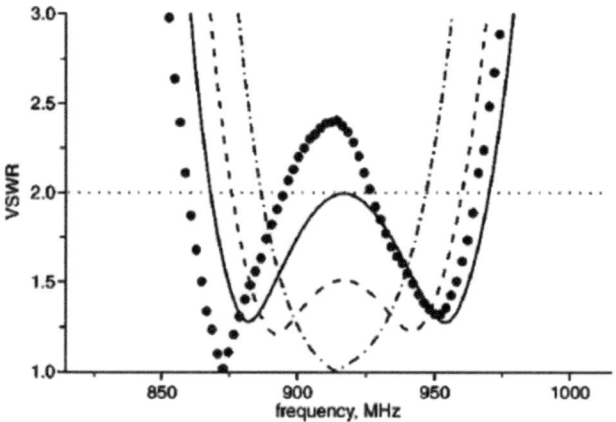

Figure 2.47. *VSWR versus frequency for different values of $R_{a,0}$:* —— $R_{a,0} = 2R_c$, − − − $R_{a,0} = 2.5R_c$, —·· — $R_{a,0} = R_c$, ● *measurements*

2.4. Combined RFID tag antenna for recipients containing liquids

The close environment of UHF RFID tags, i.e. the medium on which the inlay is fixed (plastic, metal, cardboard) and its contents (liquids, highly dielectric, metalized, etc.), degrades the radiation pattern, the antenna matching to the integrated circuit and the overall tag efficiency. This, added to the obstructions and multipath effects in indoor environments, can dramatically reduce the read-range between the reader and the tag.

The objective is to design a tag antenna attached to a plastic recipient that may be empty or filled with a liquid. A small-loop-based module excites two dipole antennas through inductive coupling. Each antenna is designed to work either for the filled or for the unfilled case at 868 MHz (UHF band).

2.4.1. *Module description*

The MuTRAK640 module manufactured by Tagsys is essentially the series connection of a small-loop antenna with a UHF RFID chip, the Impinj Monza 4. The measured chip read sensitivity and the input impedance are −14.7 dBm and 1100 Ω//2.11 pF, respectively [SAB 12]. Let $Zc = Rc + jXc$ be the series equivalent impedance of the chip load (Z_c=6.8 Ω -j86 Ω at 868 MHz). The circuit is encapsulated in a rectangular housing made up of FR4 epoxy. A key point is that the small loop dimensions yield a radiation resistance less than 1 Ω. Therefore, the loop only allows short-range reading distances (few centimeters) restricted to the reactive near-field region.

A useful feature of the module is that it can be used as a primary excitation for larger tag antennas. The idea is that a small device basically developed for short reading distances because of its low radiation resistance also performs well at distances up to 10 m when coupled with a dipole-like antenna. Typically, the magnetic field normal to the loop surface turns around the dipole located at a close proximity to the module and generates a current in the dipole. The dipole boosts the module radiation by increasing its radiation efficiency.

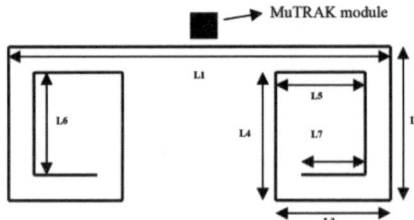

Figure 2.48. *Spiral dipole coupled to the MuTRAK module (loop+chip). Initial dimensions of the spiral dipole working in free space: L1= 40 mm, L2=20 mm, L3=13.5 mm, L4=16.5 mm, L5=10 mm, L6=13.5 mm, L7=3.5 mm, wire radius= 0.15 mm. Total length: 181 mm*

2.4.2. *Inductive coupling and antenna matching*

As seen in 2.3.1, the excitation of a dipole by an inductively-coupled loop yields the following impedance seen from the chip terminals [SON 05]:

$$Z_{antenna} = Z_{loop} + \frac{(M\omega)^2}{Z_{dipole}}$$
[2.74]

The above expression includes the mutual coupling factor M caused by the proximity between the dipole and the small loop, and the dipole impedance Z_{dipole} besides the loop impedance Z_{loop}. From [2.74], it can be observed that the series resonance of a half-wavelength dipole is transformed into a parallel resonance at the loop terminals. Equation [2.75] gives the power wave reflection coefficient of the antenna normalized to the antenna resistance [NIK 05a].

$$\Gamma = \frac{Z_c - Z_{antenna}^*}{Z_c + Z_{antenna}} \qquad [2.75]$$

2.4.3. *Antenna design*

The maximum PTC between the antenna and the chip will occur for $Z_{antenna} = Z_c^*$. Apart from the compensation of the chip capacitance by the antenna inductance, it is also necessary to keep a low antenna resistance. A dipole with spiral inductors at its ends is coupled to the encapsulated circuit MuTRAK, as shown in Figure 2.48. Spiral dipoles are compact and more efficient than zig-zag or meandered dipoles, as large gaps between the parallel segments are introduced. The dipole was made up of a thin copper wire that is fixed on a piece of paper to keep the antenna shape and rigidity. The dimensions of the dipole are given in Figure 2.49. The distance between the module edge and segment L1 is 2 mm.

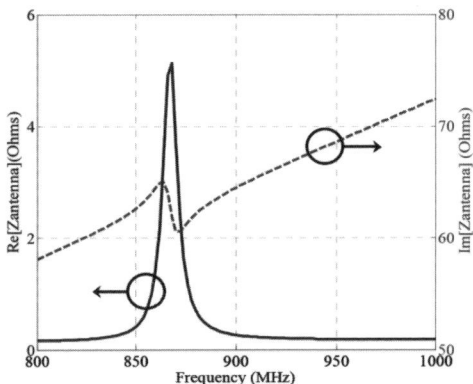

Figure 2.49. *Impedance of the loop-coupled spiral dipole*

The loop-coupled spiral dipole antenna is designed to resonate at 868 MHz. All simulations are performed with the 4NEC2 code based on the MOM. The simulated impedance is $Z_{antenna} = (5.1 + j62)\ \Omega$ for the required frequency (Figure 2.49). The estimated gain is 0 dBi. Coupling the dipole to the module introduces a parallel resonant at the chip terminals at 868 MHz as observed in Figure 2.48. As a result, the antenna impedance shows a higher radiation resistance and a flatter reactance approximately 868 GHz compared to the loop impedance. However, the reactance mean value is essentially the loop reactance that is too low at 868 MHz to exactly compensate the –86 Ω reactance of the chip. Therefore, even though the comparable antenna and chip resistances result in a Γ minimum, the total reactance (approximately –24 Ω) is large compared to the total resistance (approximately 12 Ω) which leads to Γ = 0.89. This means that the read-range which would be obtained for a perfect match is divided by $1/\sqrt{1-|\Gamma|^2} \approx 2.2$. This mismatch is inherent to the module resonance centered at 915 MHz, not to the design procedure. The following tag resonance is due to the cancellation of the total reactance and occurs above 1,000 MHz, i.e. much higher than the module resonance as the dipole coupling reduces the reactance of the isolated loop.

2.4.4. *Measurements of the initial tag*

Measurements of the dipole resonant frequency are performed with the King's shielded loop [KIN 69] used as a near-field sensor in the vicinity of the dipole. This particular probe avoids external sheet currents on the measurement cable. The dipole resonance is observed on a vector network analyzer at the King's loop terminals. The dipole length is adjusted to resonate at 868 MHz yielding the dimensions given in Figure 2.48. Then, the MuTRAK module is implemented and read-range measurements are performed with the help of the Voyantic Tagformance measurement system [OCO 09]. The tag read-range measured in the 800–900 MHz band (Figure 2.50) shows a maximum of 3.8 m at 868 MHz. This is in agreement with the estimated read-range using the modified Friis formulation given in [NIK 05a].

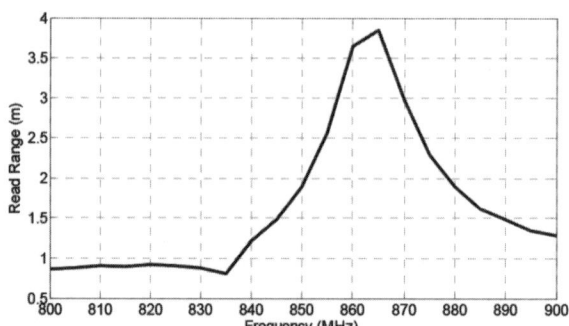

Figure 2.50. *Read-range measurements with the Voyantic setup*

2.4.5. *Measurements with an empty and filled plastic recipient*

The spiral antenna described previously is placed on an empty plastic polypropylene (ε_r = 4) recipient whose size is 23 cm × 15.5 cm × 14.5 cm. Due to the plastic wall, a 60 MHz negative shift of the resonant frequency is measured. To adjust the resonance frequency, the wire length is reduced gradually and symmetrically shortening the wire until the correct resonant frequency is obtained. A total length of 12.5 mm is finally removed from the dimensions given in Figure 2.50 (L7 = 0, L6' = 5.5 mm). The maximum read-range drops at 3.7 m.

Once the recipient is filled with fresh water ($\varepsilon_r = 80$, $\sigma = 0.1$ S/m), tag detection is not possible even at few centimeters from the reader. Using the King's shielded loop, a 550 MHz negative frequency shift of the dipole resonance is first determined. An additional 19 mm length reduction is necessary to resonate at 868 MHz (L7, L6 and L5 equal to zero and L4'=13 mm). In the presence of water, the maximum read-range is 34 cm.

2.4.6. Combined antenna

The final tag combines both proposed antennas and the MuTRAK module as shown in Figure 2.51. As the dipole resonant lengths are very different, no direct coupling between the dipoles is observed and each dipole behaves as if it was alone. Dipoles 1 and 2 resonate with the empty and filled recipient, respectively. Without water, the maximum read-range remains around 3.7 m. In the presence of water, the read-range is reduced to 31 cm, i.e. 3 cm shorter than for the single dipole. In any case, this is a strong improvement compared to the single antenna designed for the empty recipient where no detection is observed for any distance. The concept can be applied to other liquids or recipient materials as long as only two states (filled or unfilled) are considered. The tag read-range could be doubled with the MuTRAK650 version of the module based on a 3 dB more sensitive chip (Monza 5) and optimized loop dimensions.

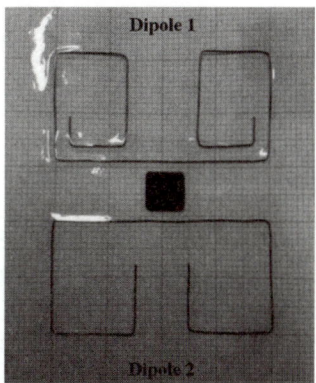

Figure 2.51. *Combined antenna structure*

2.4.7. *Discussion relative to the respect of the matching conditions*

The matching conditions must be simultaneously fulfilled for the real and imaginary parts of the input impedance of the antenna; in other words, matching is obtained when $R_{antenna} = R_c$ and $X_{antenna} = -X_c$. It is desired to study the power reflection Γ in [2.75] when only one of the two conditions is respected. To facilitate this study, Γ is plotted in Figure 2.52 as a function of $R_{antenna}$ for different values of $X_{antenna}$ with $R_{antenna}$ varying around its optimal value $R_{antenna} = 6.8\ \Omega$. In Figure 2.53, Γ is plotted as a function of $X_{antenna}$ for different values of $R_{antenna}$ with $X_{antenna}$ varying around its optimal value $X_{antenna} = 86\ \Omega$.

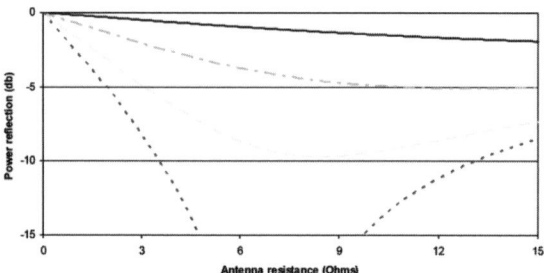

Figure 2.52. *Power reflection versus antenna reactance with Zic = (6.8–j86) Ω and for various antenna reactances : X_a=60 W ——, X_a =75 W — – – —, X_a=81 W – – –, X_a=86 W ----*

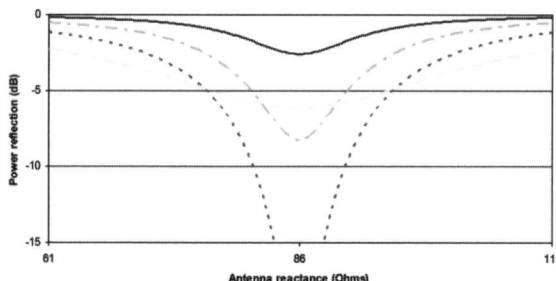

Figure 2.53. *Power reflection versus antenna reactance with Zic = (6.8–j86) Ω and for various antenna resistances R_a=1 W ——, R_a =3 W—— – – —, R_a=6.8 W – – –, R_a=20 W –*

In Figure 2.52, we first observe that once the matching condition is respected on the imaginary parts, the tolerance around the antenna optimal resistance $R_{antenna} = R_c = 6.8\ \Omega$ is large. We noted that Γ remains below -10 dB, roughly for $R_c/2 < R_{antenna} < 2\ R_c$.

Conversely, Figure 2.53 indicates that once the matching condition is respected on the real parts, the tolerance around the antenna optimal reactance $X_{antenna} = X_c = 86\ \Omega$ is small. We noted that Γ remains below -10 dB, roughly for $0.95X_c < X_{antenna} < 1.05X_c$.

We conclude that, given the typical impedances of RFID chips where series reactances are 5 to 10 times larger than series resistances, the respect of the matching condition on the imaginary part is crucial.

We also noticed that when the matching condition is far from being respected for the reactances, it is better to have $R_{antenna} >> 6.8\Omega$ as Γ tends to be less degraded than for small $R_{antenna}$ values (see curve $R_{antenna} = 20\ \Omega$ for $X_{antenna} < 70\ \Omega$ or $X_{antenna} > 100\ \Omega$ in Figure 2.53).

2.5. Tag on metal

RFID tags not only need to be mounted on non-metallic objects, but also on metallic objects. There is a strong interest from many industries (airplane, automotive, construction, etc.) in tagging metal items (airplane or automotive parts, metal containers, etc.) [ERG 07]. More generally, metal tags are usually more environmentally resistant than non-metal tags. For the tracking of high-value assets, such as industrial machinery, relatively large and complex antenna structures are acceptable to guarantee the required performance, but miniaturization approaches are needed to achieve the seamless integration of the tags with small everyday metallic items.

However, degradation in the performance of RFID tags is inescapable because of high-parasitic capacitance between the metallic surface and the antenna when tags are placed on metallic objects [UKK 05, CHO 08]. This degradation affects the radiation efficiency of the tag antenna, its radiation pattern and the input impedance. Figure 2.54 shows the effect of the gap between a metal plate and a dipole-based tag antenna. To improve the radiation efficiency and gain of tag antennas on metallic surfaces, some patch types and inverted-F types have been designed on thick and rigid substrates at high cost and with short read-range [RAO 08, KWO 05]. These antennas use shorting-pins or shorting-walls, which make their fabrication cost much higher than that of normal label-type antennas.

A key point is that metal tags require some thickness. In other words, they protrude from the mounting surfaces more or less. It is strongly desired to reduce the thickness of metal tags so as to make them more flexible in application and more aesthetic in appearance. To make these tags suitable for low-cost mass production, they should have a simple, thin structure, which is easy to fabricate from inexpensive materials.

The thicknesses of these existing designs vary from 0.8 to 2 mm. However, a thickness around 1 mm is still not small enough for tagging modern IT assets (e.g. the thickness of the latest Apple MacBook Air is less than 3 mm). Moreover, the thickness makes it difficult to curve tags for mounting on metallic cylinders (e.g. gas tank, fire extinguishers). These applications will motivate new developments on reasonably efficient solutions of ultra-low profile metal tags with thicknesses ranging from 50 μm to 300 μm.

In the following section, efficiency issues regarding low-profile patch antennas are first described. Then examples of ultra-thin and thick UHF tags are described with emphasis put on the design process, chip matching strategies and antenna performances in terms of overall efficiency.

2.5.1. Radiation efficiency of low-profile patch antennas [XI 13]

Radiation efficiency $\eta = P_{rad}/P_{in}$ is the ratio of power P_{rad} radiated into the space and the power P_{in} injected into the antenna. For patch antennas, any power consumed by the antenna other than the space-wave radiation can be classified into three loss mechanisms: dielectric loss, conductor loss and surface wave loss. For the RFID frequency bands, the substrate thicknesses are much smaller than the wavelength in the substrate so that losses in surface waves are negligible. The following compact expression for radiation efficiency [VOL 07] can eventually be determined:

$$\eta = \frac{1}{1 + Q_r \dfrac{\delta_s}{h} + Q_r \tan \delta} \qquad [2.76]$$

where σ is the metal conductivity, f is the working frequency, μ_0 is the permeability in free space and

$$\delta_s = \sqrt{\frac{1}{\pi \mu_0 f \sigma}} \qquad [2.77]$$

is the conductor skin depth. Q_r is the radiation quality factor. Its analytical expression as a function of the patch dimensions and substrate permittivity is found in [JAC 91] where it is shown that Q_r is inversely proportional to the substrate thickness and proportional to the substrate permittivity. The term $Q_r \delta_s/h$ is the power ratio of the conductor losses to the radiated power. The term $Q_r \tan \delta$ is the power ratio of the power lost in dielectric losses to the radiated power.

Note that the efficiency reduces for "bad" conductors (where σ is poor and δ_s is small) and "bad" substrates ($\tan \delta$ is high). It is also clear from [2.76] that:

– The radiation efficiency of low-profile patch antennas decreases monotonically as the substrate thickness decreases.

– When $h < \delta / \tan \delta$, the conductor loss overwhelms the dielectric loss and becomes the dominant loss mechanism.

To quantitatively demonstrate the influence of the substrate thickness, the radiation efficiency of a generic rectangular patch antenna is studied [XI 13]. The length and width of the patch antenna are 99.9 mm and 45 mm, respectively. Feeding point is at the center of one radiating edge. Calculated radiation efficiency is obtained with [2.76], while a simulated result is provided by HFSS. During simulations, a 300 mm × 200 mm metal plate is used as the ground plane, and a $\sigma = 4 \times 10^8$ S/m finite-conductivity boundary is defined to model the conductors. The substrate is a type of PET plastic, whose electric properties are $\varepsilon_r = 2.62$, $\tan \delta = 6.84 \times .0^{-3}$.

First, the radiation efficiency is studied as a function of the substrate thickness at the antenna resonance frequency (925 MHz for patch antennas studied, though it slightly varies with the substrate thickness). As shown in Figure 2.54, the agreement between the simulation and the calculation is perfect. The radiation efficiency monotonically decreases as the substrate thickness decreases.

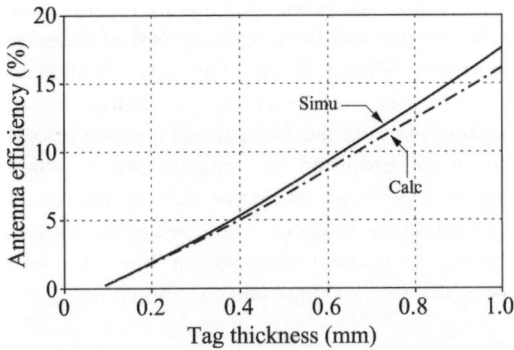

Figure 2.54. *Simulated and calculated radiation efficiency as a function of substrate thickness [XI 13]*

Next, the frequency response of the radiation efficiency is studied at three substrate thicknesses. In addition to the normal case including all loss mechanisms, two special cases where the dielectric loss and the conductor loss are excluded, respectively, are evaluated. To simulate the radiation efficiency without the dielectric loss, tanδ of the substrate is set to 0. To simulate the radiation efficiency without the conductor loss, a perfect electric-conductor (PEC) boundary is used instead of the finite conductivity boundary in HFSS. According to the simulation results shown in Figure 2.55, the frequency response of the radiation efficiency is generally quite flat, especially when all losses are included. By comparing the three cases at each substrate thickness, it is found that, at a large substrate thickness (e.g. h = 0.855 mm), the influence of the dielectric loss on the radiation efficiency is greater than that of the conductor loss. At a medium substrate thickness (e.g. h = 0.513 mm), the influences of the two loss mechanisms are comparable. At a small substrate thickness (e.g. h = 0.171 mm), the influence of the conductor loss exceeds that of the dielectric loss.

We finally conclude that a successful implementation of ultra low-profile patch antennas at extremely small thicknesses requires a good conductor. Thus, it is hardly possible to adopt conductive inks here.

2.5.2. *Ultra-thin metal tags*

This section describes two antenna designs specifically developed for thin substrates (h <1 mm) including either their own ground plane or using their support as a ground plane.

The first design is an inset-fed patch integrated with short-circuited transmission line [XI 13] where the real part and the imaginary part of the antenna impedance can be adjusted independently (Figure 2.56). The inset feeding allows the antenna resistance tuning, while the reactance is controlled by the short-circuited transmission line loading. Around the TM_{10}-mode antenna resonance frequency, the impedance behavior of the proposed tag antenna can be modeled by a lumped-element equivalent circuit as shown in Figure 2.57. In the model, the parallel RLC tank (i.e. R_a, L_a, C_a) models the inset-fed patch, while the series-connected inductor (i.e. L_t) models the short-circuited transmission line. According to the antenna model, a typical plot of the antenna impedance is shown in Figure 2.58.

The short-circuited transmission line loading is used for the adjustment of the imaginary part of the antenna impedance. According to Figure 2.58, it is a good to start trying to compensate for the chip capacitance X_{ic} with X_{max}. Using the antenna model shown in Figure 2.57, it can be opined that X_{max} satisfies the following equation:

$$X_{max} = \omega_l L_t + \frac{R_a}{2}$$

[2.78]

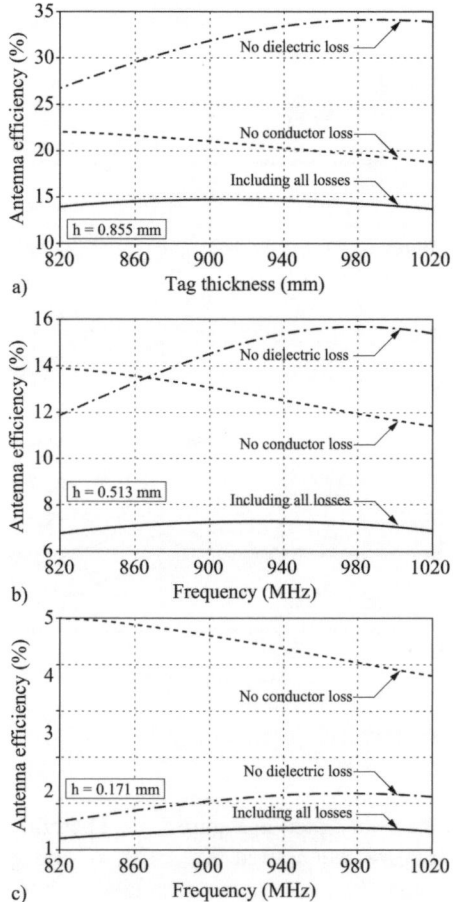

Figure 2.55. *Simulated and calculated radiation efficiency as a function of frequency for three substrate thicknesses [XI 13]*

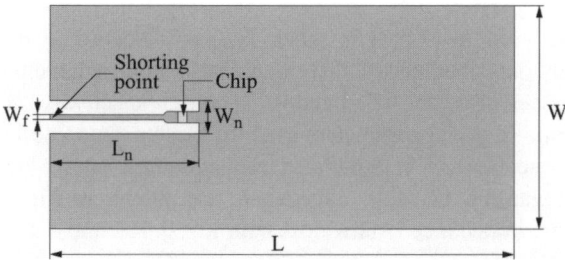

Figure 2.56. *Inset-fed patch integrated with short-circuited transmission line [XI 13]*

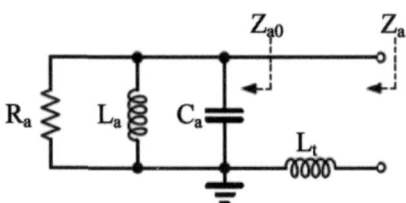

Figure 2.57. *Lumped-element model [XI 13]*

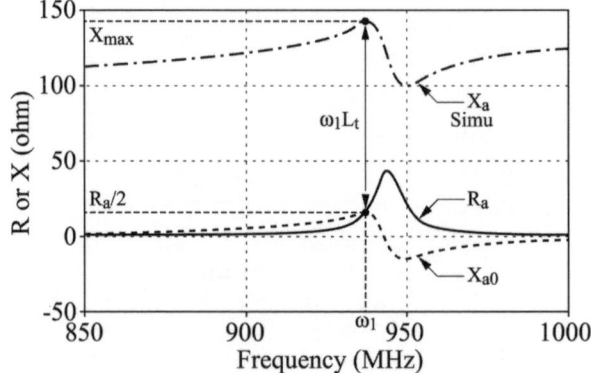

Figure 2.58. *Input impedance of the antenna [XI 13]*

Equation [2.78] is also illustrated in Figure 2.58. Therefore, L_t and R_a need to be adjusted in order to realize the desired X_{max}. Out of the geometry variables of the proposed tag antenna, L_n and W_f are two key variables to control L_t. For the adjustment of R_a, the inset feeding is used. R_a decreases as the feed approaches the center of the patch. To take care of the remaining task for the conjugate match (i.e. to make real parts of the tag antenna and the tag chip equal), a proper L_n needs to be found so that $R_a/2$ equals the chip resistance.

The tag chip used in [XI13] is Alien Higgs-3. According to the impedance matching strategy described above, three metal tag antennas have been designed for substrate thicknesses of 0.855, 0.513 and 0.171 mm, respectively. Their sizes are all 99.9 mm × 45 mm. Material parameters used for these antenna designs are the same as those used in section 2.5.1. Simulated and measured return losses (RL) of the three designs, regarding the chip impedance, are shown in Figure 2.59. Due to double-tuning, the impedance bandwidths with h = 0.513 and 0.171 mm are larger than that with h = 0.855 mm.

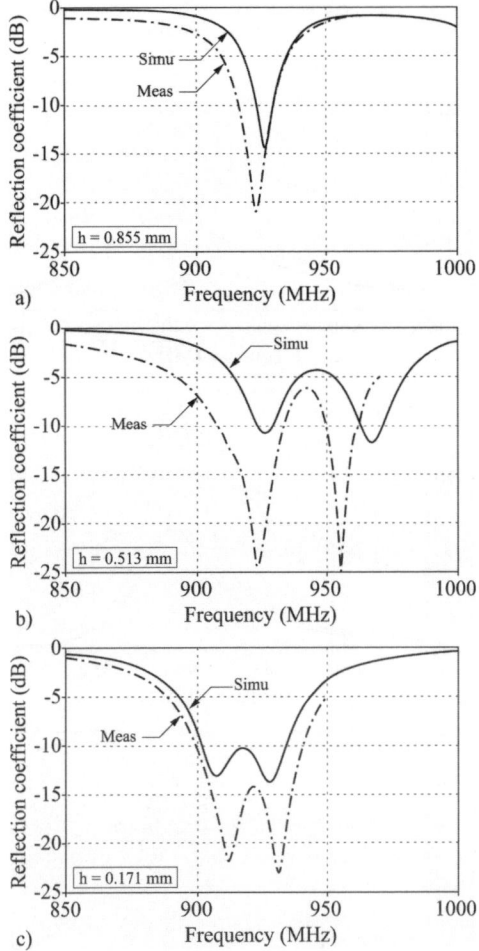

Figure 2.59. *Return loss for the three antenna thicknesses*

For "thick" substrates (say, h > 1 mm in the ultra high frequecy (UHF) RFID band), the frequency response of the radiation efficiency is stable across a very wide frequency band. Therefore, there is no need to look at the radiation efficiency bandwidth, which is much larger than the impedance bandwidth. However, for the proposed ultra-low-profile metal tag antenna, the situation changes significantly. Simulated radiation efficiency of the three tag antennas is shown in Figure 2.60. Peaks clearly emerge at the corresponding antenna resonance frequencies. Note that the antenna resonance frequency shifts a bit higher from approximately 925 MHz to about 942 MHz due to the notch in the inset-fed patch.

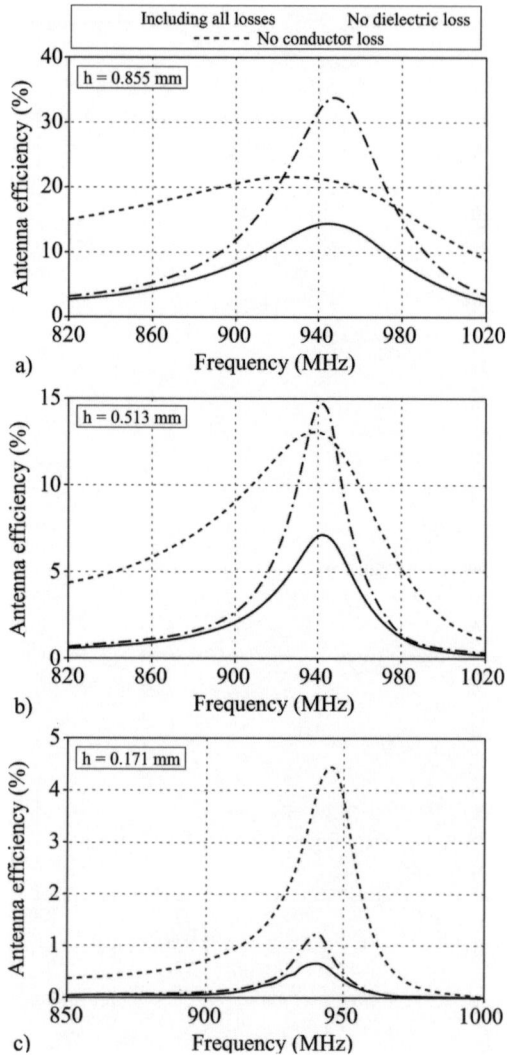

Figure 2.60. *Simulated radiation efficiency of the tag antenna designs*
as a function of frequency

For a given substrate thickness, the narrowest radiation efficiency bandwidth always happens when only accounting for the conductor loss. It implies that the conductor loss is more frequency dependent than the dielectric loss. At extremely small substrate thicknesses, the radiation efficiency bandwidth can be much narrower than the corresponding impedance bandwidth.

We conclude that when some feeding structure is added to the generic patch antenna to match the impedance of tag chips, the radiation efficiency bandwidth suffers from a significant shrinkage. As the substrate thickness decreases, the shrinkage in the radiation efficiency bandwidth becomes severer. As a result, at extremely small substrate thicknesses, it is necessary to take into account the radiation efficiency bandwidth.

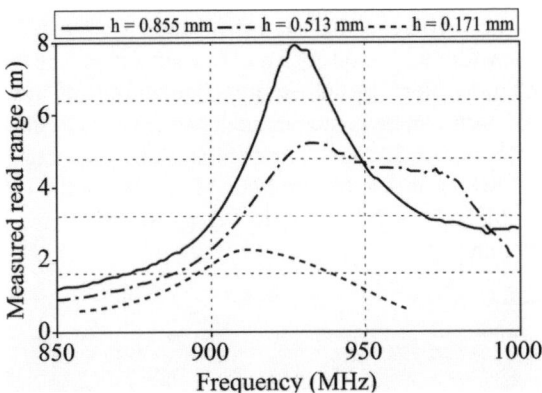

Figure 2.61. *Measured read-range of the tag prototypes*

The read-range of the tag prototypes is measured after placing tags on a metal plate of 300 mm × 200 mm. Measurement results of the read-range are shown in Figure 2.61. The maximum read-ranges with h = 0.855, 0.513 and 0.171 mm are 7.9, 5.1 and 2.2 m, respectively. It is interesting to find that the maximum read-range is approximately proportional to the substrate thickness. When h = 0.855 mm, the read-range bandwidth is determined by the impedance bandwidth. When h = 0.171 mm, the read-range bandwidth is restricted by the radiation efficiency bandwidth. In other words, for metal tag antennas implemented at extremely small substrate thicknesses, the frequency response of the read-range is shaped by the radiation efficiency rather than the power transfer coefficient.

The second design is a dual-band RFID antenna that can cover both the 866 MHz band used in Europe and the 915 MHz band used in North and South America. The configuration of the antenna is illustrated in Figure 2.62. The antenna consists of two patch arrays: A1A2 for 866 MHz and B1B2 for 915 MHz, and an ultra-thin flexible substrate polypropylene (PP) with a ground plane. Since the thickness of the flexible substrate PP (ε_r = 2.4, tan δ = 0.02) is only 0.55 mm, which generally leads to a small read-range, the arrays are designed to increase the gain of the tag antenna. The two patch arrays A1A2 and B1B2 are both symmetrical about the origin in order to conveniently connect the dual independent ports of the

Monza 4 tag chip (chip ports connect: port1 "RF1"+ "RF1" – connected to A1A2 and port2 "RF2"+ "RF2" – connected to B1B2). The two patch arrays are all short-circuited to the ground at the two ends of the substrate so that the size of patches L_p = 43.6 mm can be compacted as a quarter wavelength resonant antenna compared with a half-wavelength resonant antenna.

Although the resonant frequencies of the antenna are dominated by the length of patch L_p, the frequency difference between two bands is affected by parameters W_{p1} and W_{p2} (W_{p1} is the width of A_1 and A_2 for 866 MHz and W_{p2} is the width of B_1 and B_2 for 915 MHz). Furthermore, by adjusting the length of slots l_1, l_2 and L_s, the input impedances of dual ports can be conjugate matched easily. The actual overall size of the antenna is $L \times W = 90 \times 30$ mm^2. The optimal design parameters are as follows: W_{p1} = 14 mm (866 MHz), W_{p2} = 11 mm (915 MHz), W_{p3} = 3.4 mm, L_s = 39.5 mm, l_1 = 10.2 mm (866 MHz), l_2 = 5.5 mm (915 MHz), S_1 = 0.5 mm, S_2 =0.8 mm, S_3 = 1 mm and W_{line} = 1 mm.

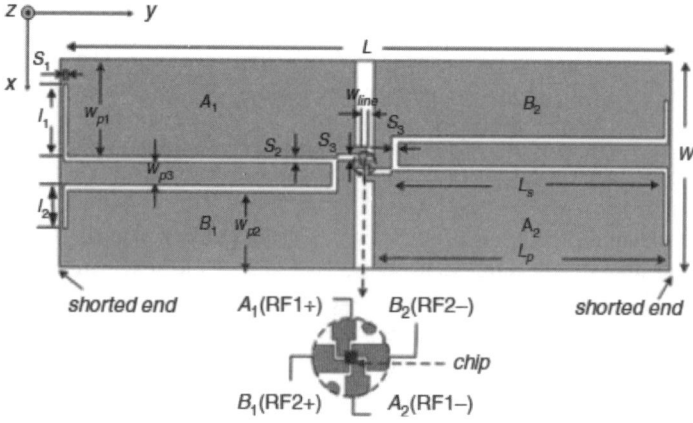

Figure 2.62. *Geometry of the dual band antenna*

Figure 2.63 shows that the Europe band (865–868 MHz) is covered by a −20 dB bandwidth of 3.5% (847 – 877 MHz) while the North America band (902–928 MHz) is covered by a −20 dB bandwidth of 5.6% (877–930 MHz). The authors claim that the −10 dB bandwidth of the proposed structure is wider than that of the conventional microstrip patch antenna. It is likely that the antenna efficiency is very low (few percents) as the calculated gain is −10 dB. The maximum read-range is 3.5 m at 866 MHz and 3.6 m at 915 MHz when the antenna is mounted on a 150 × 150 m^2 metal plate (Figure 2.64). Furthermore, the read-range curve confirms that the proposed antenna has not only a dual-band property, but also balanced read-range performance at each band.

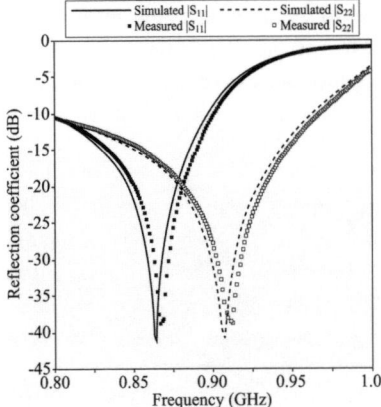

Figure 2.63. *Measured and simulated return loss of the dual-band antenna*

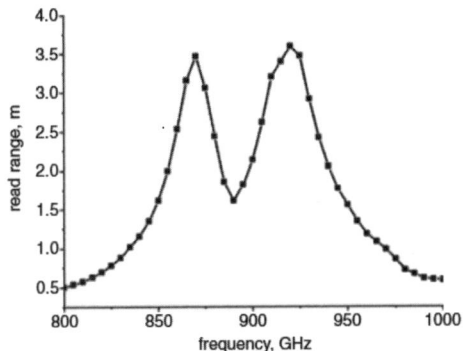

Figure 2.64. *Measured read-range of the dual-band antenna*

2.5.3. *Thick metal tags*

This section describes two topologies of modified RFID patch antennas built on low-cost thick substrates with thicknesses greater than 2 mm.

The first antenna makes use of a pair of u-shaped slots (Figure 2.65) to excite two adjacent resonant modes with similar radiation characteristics at frequencies close to each other, resulting in a wide-bandwidth patch antenna. The rectangular patch has dimensions of 40×86 mm^2, and is printed on an FR4 substrate ($\varepsilon_r = 4.4$) with a thickness of 2 mm. The microstrip feed line is deep inset into the patch to decrease the input impedance. The feed line is divided into the inset feed line (length 8 mm) and the short stub line, between which the microchip is attached. The short stub line is short-circuited to the ground plane by a via hole. A pair of U slots is etched on the patch close to and parallel to the non-radiating edges of the patch. By embedding the pair of U slots, a new resonant mode between the TM_{10} and TM_{20} modes is excited.

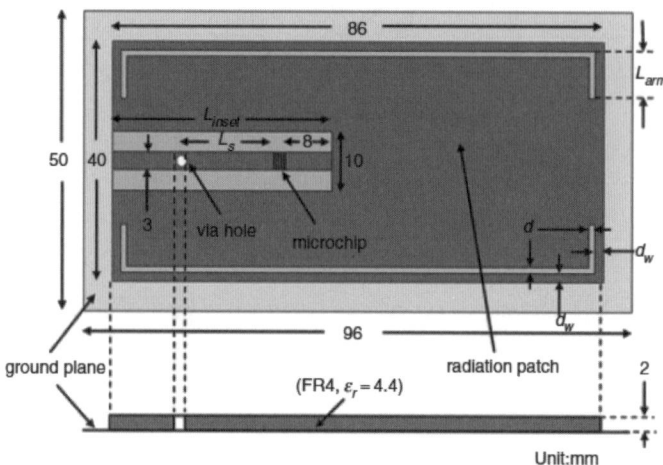

Figure 2.65. *Geometry of the dual-band antenna with a pair of U slots*

The impedance matching between the antenna and the microchip at the frequency of the TM_{10} mode can be tuned by adjusting the inset length (L_{inset}) and the distance (L_s) between the microchip and the via hole. On the other hand, the impedance matching at the frequency of the new resonant mode can be tuned by adjusting the slot arm length (L_{arm}), the slot width (d) and the distance (d_w) between the slot and the patch edge. With the increase of L_{arm}, the new resonant mode frequency decreases to the TM_{10} mode frequency. With the decrease of d and d_w, the impedance resonant amplitude of the new resonant mode could be decreased. The impedance of the microchip is $13.1-j64\ \Omega$ at 866.5 MHz and $13.5-j60\ \Omega$ at 915 MHz. When the dimensions are properly chosen as $L_{inset} = 38$ mm, $L_s = 17$ mm, $L_{arm} = 9.5$ mm, $d = 0.6$ mm and $d_w = 1$ mm, the new resonant mode is excited near the TM_{10} mode and both resonant modes have good impedance matching. In this way, a broadband characteristic can be attained for the antenna.

Figure 2.66 shows the simulated impedance of the proposed antenna with different arm lengths (L_{arm}). Note that the frequency f_1 is the TM_{10} mode and f_2 is the new resonant mode excited by the U slots between the TM_{10} and TM_{20} modes. With the increase of the arm length L_{arm}, f_2 decreases rapidly to f_1 while f_1 is slightly affected. When the arm lengths are properly chosen, the resonant mode of f_2 can be excited at the frequency that is close enough to f_1 to form a wide operating bandwidth.

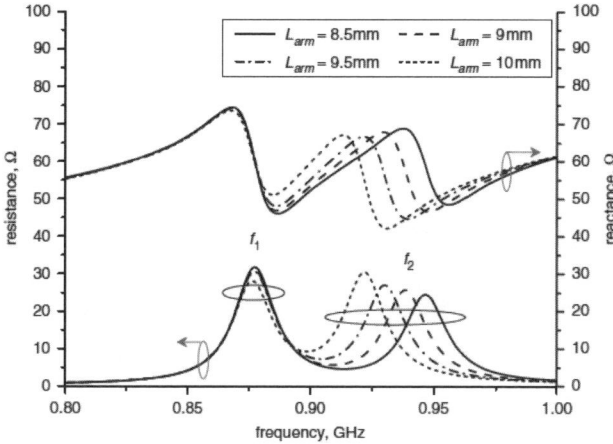

Figure 2.66. *Simulated impedance of the dual band antenna with a pair of U slots for different arm lengths*

To cover the entire operating frequency range (860–960 MHz) for passive UHF RFID worldwide, L_{arm} is taken as 9.5 mm. Figure 2.67(a) shows the simulation and measurement input impedances of the prototype antenna. The conjugate input impedance of the specified microchip is also plotted. The input resistance and reactance of the antenna vary around the conjugate value of the microchip input impedance in both f_1 and f_2 frequency ranges. Figure 2.67(b) shows the return loss of the antenna. The measured half-power bandwidth (RL = –3 dB) of the proposed antenna is 133 MHz (14.5%), from 842 to 975 MHz which covers the UHF RFID worldwide.

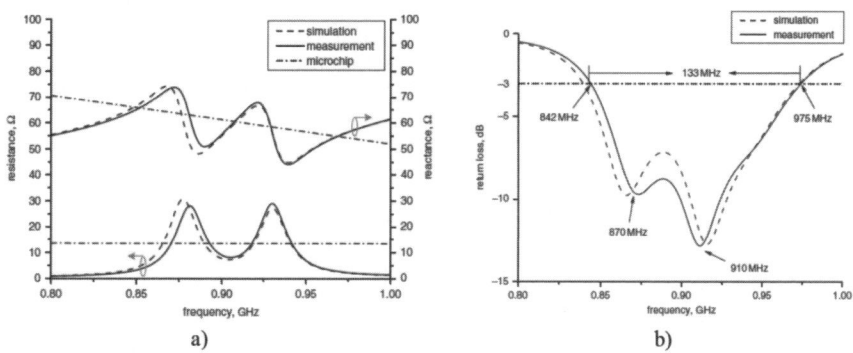

Figure 2.67. *a) Measured and simulated impedance and b) return loss of the dual-band antenna with a pair of U slots for L_{arm} = 9.5 mm*

An alternative topology proposed by Rao [RAO 08] and successfully commercialized is a patch structure with an offset coplanar tapered feeding and a virtual ground short-circuit formed by the outer rectangular ring (Figure 2.68). This tag can be used to identify and track goods and articles in various production, supply chain, or asset management scenarios, including item-level applications.

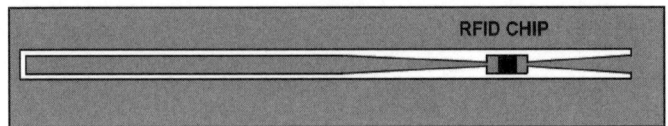

Figure 2.68. *Metal mount tag antenna (top layer antenna inlay)*

It shows two resonant frequencies (defined by complex conjugate matching to the chip impedance) to enable the wideband performance. The overall length and width of the outer rectangular ring structure determine the lower resonant frequency while the inner radiator length determines the upper resonant frequency. The tapered trace connecting the two structures provides the complex impedance matching for RFID chip.

The flexible antenna inlay was designed so that it could be placed on top of longer dielectric piece to form a large tag with high range or wrapped around smaller piece of the same dielectric to form a small tag with less range. The antenna inlay was copper etched on 2 mil flexible polyester substrate and placed on top or wrapped around the piece of polycarbonate plastic with bottom layer of conductive material, slightly longer than the dielectric itself, as shown in Figure 2.69.

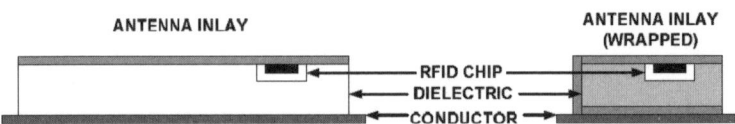

Figure 2.69. *Cross-sections of a) large and b) small versions of tag*

The final tag antenna and RFID chip were integrated and encapsulated inside rugged plastic packages capable of withstanding extreme temperatures and hazardous exposures, shown in Figure 2.70. The antenna design was slightly modified to take into account the influence of the plastic package. The large tag dimensions are $15.5 \times 3.2 \times 1$ cm^3 and the small tag dimensions are $7.9 \times 3.1 \times 1$ cm^3. Figure 2.71 presents experimental and theoretical read-range (boresight direction) in free space on plastic and metal (12×12 in). Separating two resonances makes this tag more wideband at the expense of range (bandwidth versus range tradeoff). If higher tag range in more narrow frequency band is desired, it can also be easily realized by bringing the resonances together.

Figure 2.70. *a) Large and b) small metal mount UHF RFID tags*

The two versions (large and small) of the tag are based on the same antenna inlay and deliver a global minimum read-range on metals of 25 and 10 ft accordingly. Both tags work reliably on various materials, including metal, across the worldwide UHF RFID frequency band (860–960 MHz).

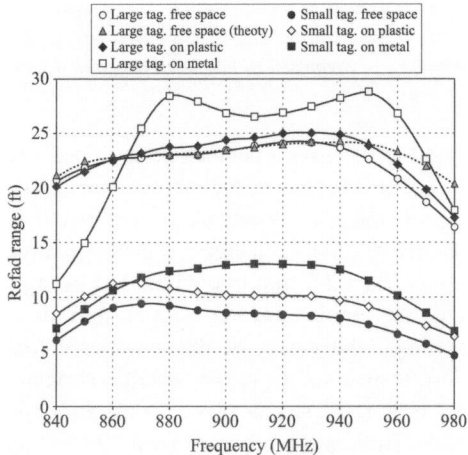

Figure 2.71. *Experimental and theoretical read-range (boresight direction) of large and small metal mount tags in free space on plastic and metal (4 W EIRP)*

2.5.4. *Improved dipole designs on metallic surfaces*

In the work described in [KOO 11], there are two primary considerations that improve the performance of the antenna on metallic objects. First, in order to increase the gain on metallic objects, an additional loop surrounds the meandered folded dipole antenna. Next, to improve matching characteristics, conjugate matching of the proposed antenna is performed in the minimum impedance variation area for the metallic object so that the change of the input impedance is small compared to free space case.

Figure 2.72(a) shows the proposed folded dipole antenna #1, which is designed without consideration for metallic objects. The antenna #1 is realized with silver paste printed on a thin flexible PET substrate with a thickness of 0.07 mm. The size for the entire substrate is 82×18 mm^2, and the line width is 1 mm.

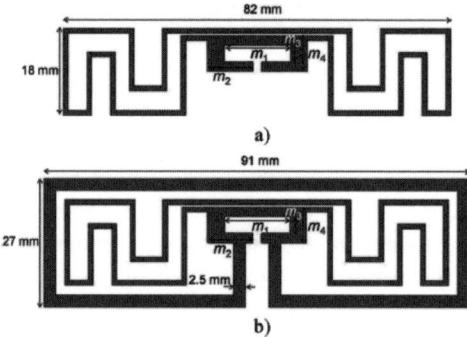

Figure 2.72. *Geometry of proposed antenna: a) antenna #1 and b) antenna #2*

The proposed method to increase antenna gain is to use a multiple folded dipole configuration [POL 07]. This purpose of technique is that antenna efficiency can be improved by increasing the radiation resistance caused by the addition of the number of folded arms. Therefore, in order to increase antenna gain on the metallic object, an additional loop, which has a 2.5 mm line width, has been placed to surround the original meandered folded dipole antenna with 1 mm line width as shown in Figure 2.72(b). To validate the effect of this additional loop, two antennas are simulated as a function of distance from the metallic surface at 910 MHz and are summarized in Table 2.6. A foam spacer ($\varepsilon_r = 1$) is used for adjusting the separation from the metallic surface. Both tags are backed by a 200×200 mm metallic surface. As seen from Table 2.6, the peak gain of antenna #2 is higher than antenna #1 in free space as well as on the metal surface. The size of antenna #2 is 91×27 mm^2.

Distance (mm)	Antenna #1 (dBi)	Antenna #2 (dBi)
1	−13.3	−11.2
2	−10.3	−4.6
3	−7.4	−2.5
Free space	0.2	1.75

Table 2.6. *Simulated results of peak gain depending on distance from the metallic surface at 910 MHz*

When the RFID tag is placed near the metallic object, the resonance moves to a lower frequency region, and then the input impedance undergoes significant frequency variation. Therefore, this change of the input impedance causes the mismatch between the chip and the antenna; so the antenna cannot deliver maximum power to the chip. To enhance the performance of RFID tags on a metallic surface, the effects of the separation from the metallic surface have been studied. Figure 2.73 illustrates the simulated impedance of the proposed antenna #2 as a function of distance between the antenna and metallic surface, and compared to the free-space case. Also, measured impedances of the IC (Alien Higgs2 in strap package) are plotted for the conjugate matching characteristic.

The IC input impedance is $(7.4 - j113)$ Ω at 910 MHz. To deliver maximum power between the IC and the antenna, the input impedance of the antenna needs to be $(7.4 + j113)$ Ω at 910 MHz. In free space, antenna #2 has an input impedance of $(6 + j110)$ Ω at 910 MHz, providing a good impedance matching to the IC input impedance. However, when antenna #2 is placed on a metallic object, significant mismatch occurs in the operating frequency band 908.5–914 MHz. As can be seen from Figure 2.73, there are large variations in the input impedance depending on the distance between the metallic surface and the antenna in the frequency region B. However, there is a relatively small variation in region A. In other words, region A shows much less impedance variation on a metallic object. Thus, if conjugate matching is performed in this region, the effect of mismatch can be minimized when it is placed on a metallic object. Therefore, in order to operate antenna #2 in region A, the resonance of antenna #2 is moved to a high frequency by reducing the entire electrical length of the antenna such that the input impedance at 910 MHz is in region A.

By adjusting the T-match, the conjugate matching between the IC and the antenna is optimized at 910 MHz in free space. The overall size of the antenna is 81×27 mm. The experimental results show that the maximum read-range of the proposed antenna is about 7.5 m in free space, and 3 m for a 1 mm separation from the metallic surface. It is interesting to note that the read-range for 10 mm separation is 7.32 m, which is similar to that of free space.

In addition, the proposed antenna is compared to a commercial RFID label-typed tag (Alien M-Tag) also equipped with the Alien Higgs 2 chip and fabricated out of copper. Also, the overall size of the commercial tag is similar to the proposed antenna. Figure 2.74 shows a comparison between the commercial RFID label-typed tag and the proposed antenna below 5 mm separation. The proposed antenna reveals a much longer read-range than the commercial RFID antenna. For example, the read-range of the proposed antenna is about three times longer than that of the Alien M-tag when it is placed 1 mm away from the metallic surface.

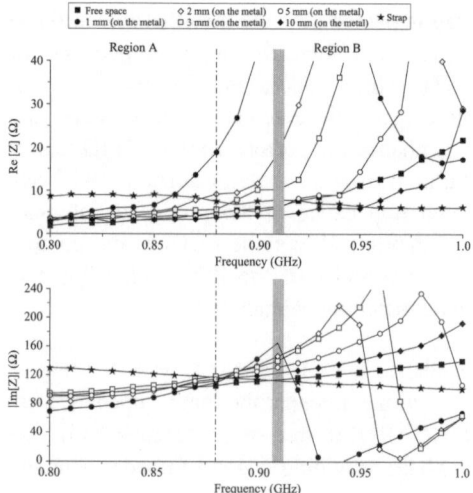

Figure 2.73. *Simulated input impedance as a function of distance from the metallic surface*

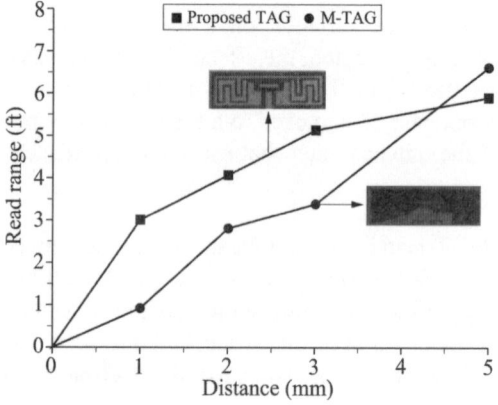

Figure 2.74. *Read-range comparison against the commercial RFID tag*

2.6. Bibliography

[BAL 05] BALANIS C.A., *Antenna Theory: Analysis and Design*, 3rd ed., John Wiley & Sons, 2005.

[BES 05] BEST R., "Low Q electrically small linear and elliptical polarized spherical dipole antennas," *IEEE Transactions on Antennas and Propagation*, vol. 53, no. 3, pp. 1047–1053, March 2005.

[CHO 08] CHO C., CHOO H., PARK I., "Design of planar RFID tag antenna for metallic objects", *Electronics Letters*, vol. 44, pp. 175–177, 2008.

[CHO 07] CHO C., KIM D., CHOO H., *et al.*, "Effect of the substrate, metal-line and surface material on the performance of RFID tag antenna", *IEEE Antennas and Propagation Society International Symposium*, pp. 1761–1764, 9–15 June 2007.

[DEA 09] DEAVOURS D.D., "Analysis and design of wideband passive UHF RFID tags using a circuit model", *IEEE RFID Conference*, Orlando, FL, April 2009.

[DEA 10] DEAVOURS D., "UHF RFID Antennas", *RFID Systems: Research Trends and Challenges*, Chapter 3, John Wiley & Sons, 2010.

[DOB 07] DOBKIN D.M., *The RF in RFID: Passive UHF RFID in Practice Burlington*, Newnes, 2007.

[DOB 12] DOBKIN D.M., *The RF in RFID: UHF RFID in Practice*, Newnes, 2012.

[DU 13] DU G.H., TANG T., DENG Y., "Dual-band metal skin UHF RFID tag antenna", *Electronics Letters*, vol. 14, pp. 858–860, 4 July 2013.

[END 00] ENDO T., SUNAHARA Y., SATOH S., *et al.*, "Resonant frequency and radiation efficiency of meander line antennas," *Electronics and Communications in Japan, Part 2 (Electronics)*, vol. 83, pp. 52–58, 2000.

[ERG 07] ERGEN E., AKINCI B., EAST B., *et al.*, "Tracking components and maintenance history within a facility utilizing radio frequency identification technology", *Journal of Computing in Civil Engineering*, vol. 21, no. 1, pp. 11–20, January 2007.

[GRO 04] GROVER F.W., *Inductance Calculations*, Dover Publications, Mineola, 2004.

[HAZ 11] HAZDRA P., CAPEK M., EICHLER J., "Radiation Q-factors of thin-wire dipole arrangements", *IEEE Antennas and Wireless Propagation Letters*, vol. 10, pp. 556–560, 2011.

[HU 09] HU Z., COLE P.H., ZHANG L., "A method for calculating the resonant frequency of meander-line dipole antenna", *Industrial Electronics and Applications (ICIEA) Conference*, pp. 1783–1786, Xian, China, May 2009.

[JAC 91] JACKSON D.R., ALEXOPOULOS N.G., "Simple approximate formulas for input resistance, bandwidth, and efficiency of a resonant rectangular patch", *Transactions on Antennas and Propagation*, vol. 39, no. 3, pp. 407–410, 1991.

[KIN 69] KING R.W.P., "The loop antenna for transmission and reception", *Antenna Theory*, Chapter 11, McGraw-Hill, NY, p. 478, 1969.

[KOO 11] KOO T., KIM D., RYU J., *et al.*, "Design of a label-typed UHF RFID tag antenna for metallic objects", *IEEE Antennas and Wireless Propagation Letters*, vol. 10, pp. 1010–1014, 2011.

[KWO 05] KWON H., LEE B., "Compact slotted planar inverted-F RFID tag mountable on metallic objects", *Electronics Letters*, vol. 44, pp. 1308–1310, 2005.

[LOP 07] LOPEZ A.R., "Wheeler and Fano impedance matching", *IEEE Antennas and Propagation Magazine*, vol. 49, no. 4, pp. 116–119, 2007.

[MAR 08] MAROCCO G., "The art of UHF RFID antenna design: impedance matching and size-reduction techniques", *IEEE Antennas and Propagation Magazine*, vol. 50, no. 1, pp. 66–79, February 2008.

[MCL 96] MCLEAN J.S., "A re-examination of the fundamental limits on the radiation Q of electrically small antennas", *IEEE Transactions on Antennas and Propagation*, vol. 44, no. 5, pp. 672–676, May 1996.

[MCD 12] MCDONALD K.T., Reactance of small antennas, Joseph Henry Laboratories, Princeton University, Princeton, NJ, 2012.

[MO 08] MO L., ZHANG H., ZHOU H., "Broadband UHF RFID tag antenna with a pair of U slots mountable on metallic objects", *Electronics Letters*, vol. 44, no. 20, pp. 1173–1174, 25 September 2008.

[NIK 05a] NIKITIN P.V., RAO K.V.S., LAM S.F., *et al.*, "Theory power reflection coefficient analysis for complex impedances in RFID tag design", *IEEE Transactions on Microwave Theory and Techniques*, vol. MTT-53, no. 5, pp. 2721–2724, September 2005.

[NIK 05b] NIKITIN P.V., LAM S., RAO K.V.S., "Low cost silver ink RFID tag antennas", *IEEE Antenna and Propagation Society International Symposium*, vol. 2B, pp. 353–356, July 2005.

[OCO 09] O'CONNOR M.C., "Voyantic helps companies put RFID tags to the test", *RFID Journal*, June 18, 2009. Available at: http://www.rfidjournal.com/article/view/4983.

[POL 07] POLIVKA M., AMBEMOU E., "Electrically shortened wire RFID antennas closely spaced to a PEC," *Proceedings of the 2nd European Conference on Antennas and Propagation*, pp. 1–5, November 2007.

[RAO 05] RAO K.V.S., NIKITIN P.V., LAM S., "Antenna design for UHF RFID tags: a review and a practical application", *IEEE Transactions on Antennas and Propagation*, vol. 53, no. 12, pp. 3870–3876, December 2005.

[RAO 08] RAO K.V.S., LAM S.F., NIKITIN P.V., "Wideband metal mount UHF RFID tag", *Proceedings of Antennas and Propagation Society International Symposium 2008*, San Diego, CA, pp. 1–4, 5–11 July 2008.

[SAB 12] SABA R., DELERUYELLE T., ALARCON J., *et al.*, "A resistant textile tag antenna for RFID UHF frequency band", *IEEE Conference on RFID Technology and Applications*, Nice, France, pp. 203–207, 5–7 November 2012.

[STU 12] STUTZMAN W.L., THIELE G.A., *Antenna Theory and Design*, 3rd ed., Wiley, June 2012.

[SON 05] SON H.-W., PYO C.-S., "Design of RFID tag antennas using an inductively coupled feed", *Electronics Letters*, vol. 41, no. 18, p. 2, September 2005.

[TAI 07] TAI C.-T., LONG S.A., "Dipoles and monopoles", in *Antenna Engineering Handbook*, 4th ed., JOHN L. VOLAKIS, ed., McGraw Hill, 2007.

[UKK 05] UKKONEN L., SYDANHEIMO L., KIVIKOSKI M., "Effects of metallic plate size on the performance of microstrip patch-type tag antennas for passive RFID", *IEEE Antennas and Wireless Propagation Letters*, vol. 4, pp. 410–413, 2005.

[VOL 07] VOLAKIS J., *Antenna Engineering Handbook*, 4th ed., McGraw-Hill, 2007.

[XI 11] XI J., YE T., "Wideband and material-insensitive RFID tag antenna design utilizing double-tuning technique", *International Symposium on Antennas and Propagation (APSURSI)*, Spokane, Washington, DC, pp. 545—548, 3–8, July 2011.

[XI 13] XI J., YE T.T., "Ultra low-profile metal tag antenna design with an emphasis on radiation efficiency", *2013 IEEE International Conference on RFID (RFID '13)*, Orlando, FL, pp. 42–49, 30 April–02 May 2013.

[YAG 05] YAGHJIAN A.D., BEST S.R., "Impedance, bandwidth and Q of antenna", *IEEE Transactions on Antennas Propagation*, vol. 53, no. 4, pp. 1298–1324, April 2005.

3

The Backscattering Technique and its Application

Before starting …

Today backscattering and shunt regulator techniques in RFID are not new … but their mutual incidences are not very well known. So, to give to the reader a good and complete overview about this complex subject, a significant part of the content of this chapter is issued from the already existing Dominique Paret books: *RFID en Ultra et Super Hautes Fréquences UHF-SHF: Théorie et mise en oeuvre*, Dunod, 2005/2008, and the English version *RFID at Ultra and Super High Frequencies: Theory and Application*, John Wiley, 2009.

As shown in previous chapters, the power P_s reflected or reradiated by the tag (which is dependent on the value of the power flux density s) can be received and detected by the receiving antenna of the base station and can thus act as a signal informing the base station of whether or not an object or tag is present in the electromagnetic field.

Also, while the tag is illuminated, and regardless of whether it is remotely powered or locally battery assisted, provided that it has been designed to respond accurately via a specific modulation, it is called "backscattering modulation", which will be described in detail later on.

Therefore, it is useful to analyze the way in which the radar cross section (RCS) is varied or modulated, and to define the extent of its variation – the quantity $\Delta\sigma_{es}$ – as a function of a possible coding and a specific modulation, which will enable us to determine its ability to be understood correctly by the base station. We will then proceed to determine what its merit factor is, or ought to be.

3.1. Backscattering principle of communication by between-base station and tag

As a general rule, allowing for exceptions (and there are some), the communication model followed by standard RFID systems used at UHF and SHF is based on the reader talk first (RTF) principle, using half-duplex mode (an alternating link between the base station and the tag).

The stages of transmission are summarized in Figure 3.1.

3.1.1. *The forward link: communication from the base station to the tag*

During the first phase, known as the forward link of the half-duplex, the base station transmits the carrier frequency to remotely power the batteryless tag. During this phase, the carrier frequency is modulated (in amplitude shift keying (ASK) mode, for example) for the transmission of the command and interrogation codes to the tag.

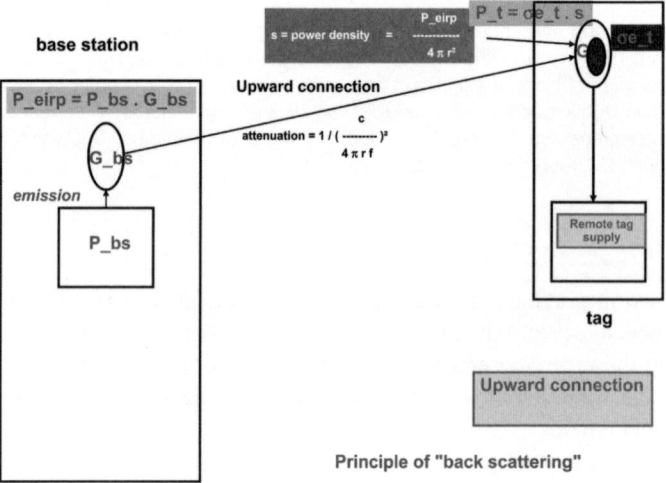

Figure 3.1. *Principle of backscattering: forward link*

Note that, during this phase, the tag illuminated by the incident electromagnetic wave may either absorb the power that it receives or reradiate some of it, depending on the state of its antenna/load impedance matching. Generally, for remotely powered tags, the tag is made to absorb the maximum possible power (i.e. with no standing waves) during this phase, in order to provide the best possible remote power supply and thus achieve the highest possible operating distance. However, it may reradiate during this phase, according to its "structural" aspect.

To sum up, during the forward link we have:

– conjugate impedance matching (in remotely powered tags);

– maximum power transmitted to the load (remote power supply);

– reradiation of the "structural" type.

3.1.2. *The return link: communication from the tag to the base station*

During the second phase of the half-duplex, called the "return link", with transmission from the tag, the base station initially supplies or maintains the sustained pure (unmodulated) carrier frequency to provide a physical support for the tag's response following the preceding interrogation commands. During this return link phase, two operating subphases may be present, depending on the binary information to be transmitted by the tag to the base station:

a) Either the transmission of no useful information or the transmission of a logical "1": note that these are often identical in physical terms, since they correspond to the same phenomenon as that described in the previous section concerning the forward link.

b) Alternatively, the transmission of a logical "0": in this case, the tag's electronic circuit modulates the value of the load impedance $Z_l = R_l + X_l$ of the tag antenna at the rate of an on-off keying (OOK) modulation corresponding to the logical data to be transmitted. Thus, at the tag, there will be an impedance mismatch between the source (the tag antenna) and its load, leading to the appearance of standing waves, and therefore to a new effective RCS and a variation of the RCS area, which will immediately modify the amount of power reradiated in a different way from that described in the previous section.

NOTES.–

The receiving part of the base station examines the content of the return path (the return link) during subphases (a) and (b) only (Figure 3.2.).

To make this phenomenon easier to understand, the explanations above relate to the case of simple bit coding (NRZ) of the return link. This bit coding is often different (Manchester, BPSK, FM0, etc.).

To sum up, during the return link we have:

– deliberate mismatching of the antenna load impedance;

– a load power mismatch factor q;

– a change in the RCS of the tag;

– the reradiation of a different power levels by the tag, signifying the presence of a bit of opposite value.

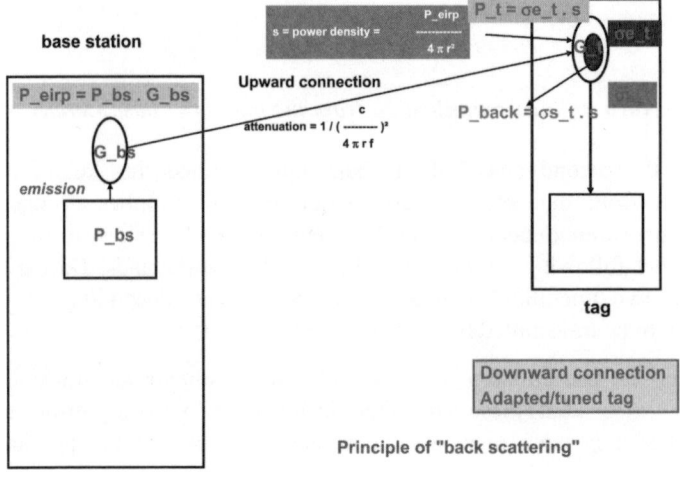

a)

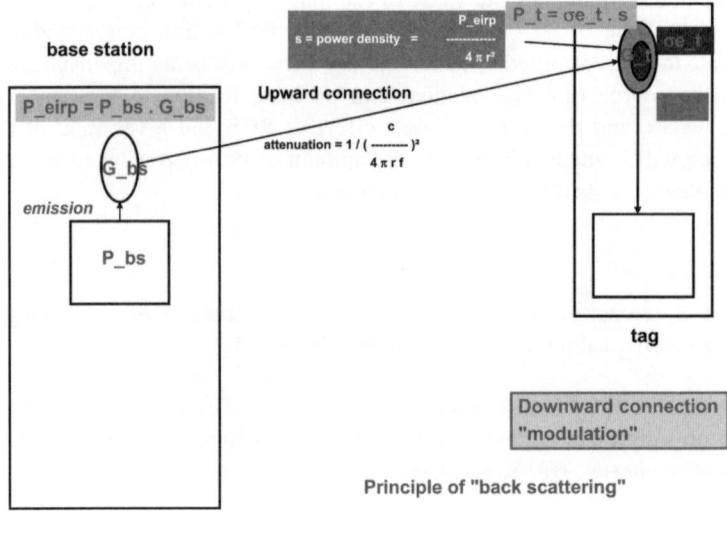

b)

Figure 3.2. *Principle of backscattering a) return link, tag matched/tuned; b) "modulation" return link*

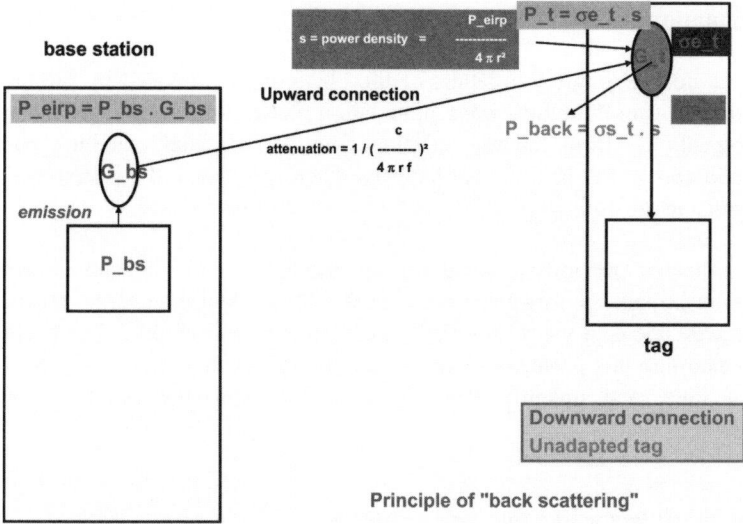

Figure 3.3. *Principle of backscattering: return link, tag mismatched*

The communication concept of the return link as outlined above is the main basis of UHF and SHF RFID systems operating in a mode of detecting the value of the reradiated/scattered return wave, known as the backscattering modulation mode.

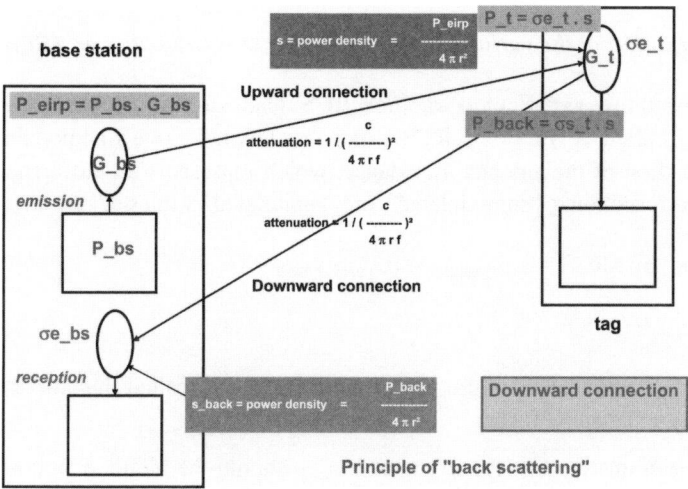

Figure 3.4. *Principle of backscattering: return link*

... continuation

Let us now examine the details of the phenomena produced by the tag in the return link, during the return wave modulation phase, for example the transmission of a logical "0" from the tag, starting from the optimal matching conditions mentioned above. Let us see what happens when we change the load impedance to values outside the optimal matching conditions mentioned above.

As indicated previously, the deliberate modulation of the load impedance Z_l leads to an impedance mismatch between the source and the charge, on one hand, and the presence of a wave reradiation phenomenon, on the other hand. Therefore, we can examine this problem of impedance modification in terms of a "distributed constant line" and quantify this mismatch by using the reflection factor to calculate it.

3.2. The merit factor of a tag, $\Delta\sigma_{es}$ or ΔRCS

As shown above, in RFID applications, which operate on the backscattering principle and which are therefore based on the principle of modulating the wave reradiated by the tag, the effective RCS of the tag σ_{es} varies when the impedance of the tag antenna is deliberately modified by changing the value of its resistive and/or capacitive part.

3.2.1. *Definition of the variation of the radar cross section, σe s or ΔRCS*

The resulting variation of σ_{es} or RCS leads to the appearance of a new parameter, called $\Delta\,\sigma es$ or ΔRCS = σe s modulated $-$ σe s unmodulated due to the modulation of the antenna impedance, which represents the difference between the two corresponding "unmodulated" and "modulated" values of σ_{es}.

$$\Delta\sigma_{es} \text{ or } \Delta\text{RCS} = \sigma_{\text{e s modulated}} - \sigma_{\text{e s unmodulated}}$$

NOTES.–

1) In the technical literature, $\Delta\sigma_{es}$ or ΔRCS is also called the "merit figure" or "merit factor" of an RFID tag.

2) Theoretically, the merit factor $\Delta\sigma_{es}$ should be defined not as a scalar magnitude but as a vector value (i.e. complex variable) and therefore it includes amplitude and direction (or modulus and phase), allowing for all the complex impedances involved.

This is because, depending on the different modulation states (particularly in the case of BPSK or QPSK), the values of the reradiated scattered power in watts may be equal while the phases of the scattered waves are different, and the difference of the scalar quantities $\Delta\sigma_{e\,s}$ between two states of modulation may be zero. To avoid this, the demodulator in the receiving part of the base station must allow for this possibility and must therefore be capable of demodulating either the amplitude or the phase of the received signal equally well.

Let us now take an overview of its variations as a function of the different parameters.

3.2.2. Estimation of $\Delta\sigma_{e\,s}$ as a function of $\Delta\Gamma$

In an RFID application, we do not know in advance what the initial "unmodulated" position of the tag corresponds to in physical terms (is it matched, nearly matched or completely unmatched?), or what the corresponding value of $\Gamma_1 = \Gamma_{\text{non modulated}}$ will be. To make matters clear, the "unmodulated" physical state depends on the principles adopted for the design of the application. For example, the system designer may decide that the tag should be of the battery assisted type (having an incorporated battery, but still of the passive communication type return link) because the application requires operation at very long range, in which case the tag antenna does not necessarily have to be matched in advance in the "unmodulated" position to recover the maximum energy, since a battery is provided in the tag.

By contrast with many currently available books on this subject, I will cover every possible kind of application by stating, without any prior assumptions, that

$\Gamma_1 = \Gamma_{\text{non modulated}}$ (initial value in the unmodulated position)

$\Gamma_2 = \Gamma_{\text{modulated}}$ (the value when the load impedance is switched, i.e. in the modulated position)

$$\Delta\Gamma = \Gamma_2 - \Gamma_1 = \Gamma_{\text{modulated}} - \Gamma_{\text{non modulated}}$$

Using the general equation, we can write:

$-\sigma_{e\,s} = \sigma_{e\,s\;\text{structural}} + \sigma_{e\,s\;\text{antenna mode}}$ = fixed part + variable part (positive or negative) as a function of the load R_l

$-\sigma_{e\,s} = \sigma_{e\,s\;\text{structural}} + [\{(1 - \Gamma)^2 - 1\}\sigma_{e\,s\;\text{structural}}]$

$-\sigma_{e\,s} = (1 - \Gamma)^2\,\sigma_{e\,s\;\text{structural}}$

Because the structure of the equation $\sigma_{e\,s} = \sigma_{e\,s\,structural} + \sigma_{e\,s\,antenna\,mode}$ has a fixed part and a variable part, the difference between the "modulated" and "unmodulated" states is simply the algebraic difference between the two values of the variable part of the equation $\sigma_{e\,s\,antenna\,mode}$, namely $\sigma_{e\,s\,antenna\,mode\,mod}$ and $\sigma_{e\,s\,antenna\,mode\,non\,mod}$ (Figure 3.5).

$$\Delta\sigma_{e\,s} = \Delta RCS \text{ of the tag} = \sigma_{e\,s\,antenna\,mode\,mod} - \sigma_{e\,s\,antenna\,mode\,non\,mod}$$

Now let us calculate the corresponding values of $\sigma_{e\,s1}$ and $\sigma_{e\,s2}$. We obtain:

– in the "unmodulated" phase: $\sigma_{e\,s1} = \sigma_{e\,s\,structural} + [\{(1 - \Gamma_1)^2 - 1\}\sigma_{e\,s\,structural}]$;

– in the "modulated" phase: $\sigma_{e\,s2} = \sigma_{e\,s\,structural} + ([(1 - \Gamma_2)^2 - 1]\sigma_{e\,s\,structural}$.

After reduction, the merit factor becomes:

– $\Delta\sigma_{e\,s} = \sigma_{e\,s2} - \sigma_{e\,s1}$;

– $\Delta\sigma_{e\,s} = \Delta\Gamma\,[-2 + (\Gamma_2 + \Gamma_1)]\sigma_{e\,s\,structural}$.

Finally, by replacing Γ_2 in this equation by its value $(\Delta\Gamma + \Gamma_1)$, we obtain:

$$\Delta\sigma_{e\,s} = \Delta\Gamma\,[\Delta\Gamma + 2(\Gamma_1 - 1)]\sigma_{e\,s\,structural}$$

$$\Delta\sigma_{e\,s} = \Delta\Gamma\,[\Delta\Gamma + 2(\Gamma_1 - 1)]\,\frac{\lambda^2 G_{ant\,t}^2}{4\pi} = f(\Delta\Gamma, \Gamma_1) \text{ in m}^2$$

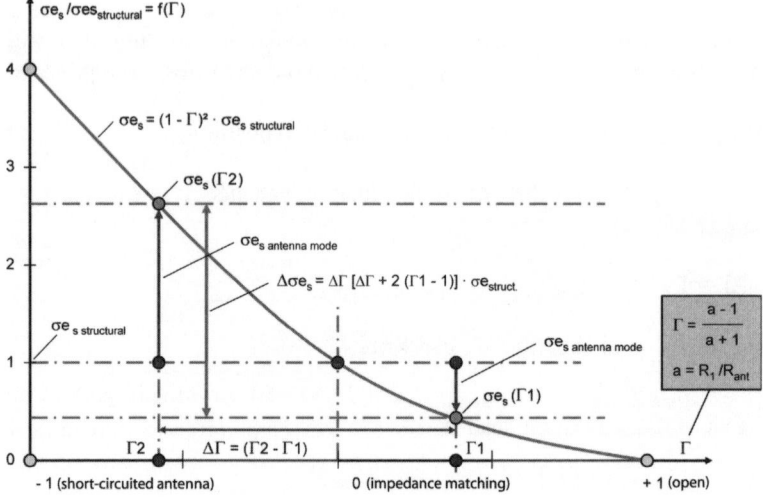

Figure 3.5. *Calculation and variations of $\Delta\sigma_{e\,s} = \Delta RCS$*

It is very important to note that the function $\Delta\sigma_{e\,s} = \Delta RCS$ is simultaneously dependent on two elements: the variable $\Delta\Gamma$ and a parameter representing the initial value of Γ_1.

3.2.3. The variation $\Delta\sigma_{e\,s} = f(\Delta\Gamma,\Gamma_1)$

Let us return to the general equation for $\Delta\sigma_{e\,s}$ found above:

$$\Delta\sigma_{e\,s} = \Delta\Gamma\,[\Delta\Gamma + 2(\Gamma_1 - 1)]\sigma_{e\,s\,\text{structural}}$$

First, we should note that the value of $\Delta\sigma_{e\,s}$ is a function of two intercorrelated variables, and for simplicity we will examine the case where $\Delta\sigma_{e\,s} = f(\Delta\Gamma)$ with $\Gamma_1 =$ constant.

Let us start with an observation. Theoretically, the maximum variation of $\Delta\Gamma$ is limited to the range of 0–2 inclusive, because the value of Γ can vary from –1 to +1. In practice, in the case of numerous RFID applications of the remotely powered type, this range of variation will be smaller (from 0 to – 1) because of the need to recover some of the incident energy to supply the remotely powered tags (see the list in Table 3.1).

Γ_1	Γ_2	$\Delta\Gamma = \Gamma_2 - \Gamma_1$	$\Delta\sigma_{e\,s} = \Delta\Gamma\,[\Delta\Gamma + 2(\Gamma_1 - 1)]\sigma_{e\,\text{structural}}$
+1	0	–1	$= -1[-1 + 2 \times (1 - 1)] = -1 \times \sigma_{e\,\text{structural}}$
+1	–1	–2	$= -2[-2 + 2 \times (1 - 1)] = +4 \times \sigma_{e\,\text{structural}}$
0	–1	–1	$= -1[-1 + 2 \times (0 - 1)] = +3 \times \sigma_{e\,\text{structural}}$
0	–1	+1	$= +1[1 + 2 \times (0 - 1)] = -1 \times \sigma_{e\,\text{structural}}$ (as expected)
–1	0	+1	$= +1\,[+1 + 2 \times (-1 - 1)] = -3 \times \sigma_{e\,\text{structural}}$
–1	+1	+2	$= 2\,[2 + 2 \times (-1 - 1)] = -4 \times \sigma_{e\,\text{structural}}$

Table 3.1.

Now let us take a closer look at the two very common cases shown in the first two columns in Table 3.1, which represent the great majority of conventional RFID applications.

– The usual case of "remotely powered" applications, $\Gamma_1 = 0$ and $\Gamma_2 = x$

Large number of RFID applications operate in "remotely powered" mode and therefore start from the initial unmodulated position with conjugate matching, i.e. $R_l = R_{ant\ t}$ and $a = 1$, in other words, $\Gamma_1 = 0$, and then switches to a value of R_l different from that of $R_{ant\ t}$, giving rise to a new value, Γ_2. In this case alone, $\Delta\Gamma = \Gamma_2$, and the above equation is simplified and reduced to the form:

$$\Delta\sigma_{e\,s} = \Gamma_2\,[-2 + \Gamma_2]\sigma_{e\ structural};$$

$$\Delta\sigma_{e\,s} = \Gamma_2\,(\Gamma_2 + 2)\sigma_{e\ structural};$$

$$\Delta\sigma_{e\,s} = [\Gamma_{mod}^2 - 2\Gamma_{mod}^2]\,\sigma_{e\ structural},\ \text{as mentioned above.}$$

Table 3.2 shows examples of values of $\Delta\sigma_{e\,s}$.

R_l	a	Γ_2	$\Gamma_2\,(\Gamma_2 - 2)$	$\Delta\sigma_{e\,s}$
0	0	−1	−3	$3 \times \sigma_{e\,s\ structural}$
R_{ant}	1	−0	−0	0
∞	∞	+1	−1	$-1 \times \sigma_{e\,s\ structural}$

Table 3.2.

Let us look at two subcases that frequently arise when $\Gamma_1 = 0$.

– Subcase 1

R_l is often switched all the way from the conjugate matching condition, $R_l = R_{ant\ t}$, $a = 1$, to $R_l = 0$ (load short-circuited) in UHF RFID, in order to maximize the variation of the value of $\Delta\sigma_{e\,s}$. This simultaneously gives rise to two conditions, $a = 0$ and $\Gamma = -1$, and therefore:

$$\Delta\sigma_{e\,s} = \sigma_{e\,s} - \sigma_{e\ structural} = (4 - 1)\,\sigma_{e\ structural} = 3\sigma_{e\ structural}$$

– Subcase 2

In RFID, it is sometimes desirable to reach a compromise between the variation of the RCS and the power consumption of the tag in order to optimize the operating range of the system, and therefore R_l is made to change only very slightly about the matching value $R_{ant\ t}$. In this case only, $a = R_l/R_{ant\ t}$, i.e. close to or substantially equal to 1, and that Γ_2 is (very) close to 0 (matching). In this case, Γ_2^2 will be small with respect to $-2\Gamma_2$. Consequently, in this particular case only,

$$\Delta\sigma_{e\,s} = \sigma_{e\,s\ antenna\ mode} \cong -2\Gamma_2\sigma_{e\ structural}$$

which fits very well on the curve (the slope of the curve where $a = 1$, i.e. $\Gamma = 0$).

– The normal case of "battery assisted" applications, $\Gamma_1 = +1$ and $\Gamma_2 = -1$

If there is no need to provide a remote power supply to the tag, because the applications concerned use battery-assisted tags, then our aim will clearly be to benefit from the wider variation of $\Delta\sigma_{e\,s}$ by switching the load condition from "fully open" to completely "short-circuited". If the initial value of Γ in the unmodulated condition, Γ_1, is equal to 1 (i.e. there is an open load) – and only in this case – the above equation becomes:

$$\Delta\mathrm{RCS} = \frac{\lambda^2}{4\pi} G_{\text{tag}}^2 \, \Delta\Gamma^2$$

and clearly we can write

$$\Delta\mathrm{RCS} = \frac{\lambda^2}{4\pi} G_{\text{tag}}^2 \, (\Gamma_2 - \Gamma_1)^2$$

an equation which appears in many books and documents in this field ... but unfortunately with no indication of the limits of its validity (battery-assisted tags only). So we should be careful with unforeseen results. Table 3.3 shows the values for $\Gamma_1 = 1$.

Γ_1	Γ_2	$\Delta\Gamma$	$\Delta\sigma_{e\,sx} = \Delta\Gamma\,[\Delta\Gamma + 2(\Gamma_1 - 1)]\sigma_{e\text{ structural}}$
$+1$	-0	-1	$= -1\,[-1 + 2 \times (1 - 1)] = +1 \times \sigma_{e\text{ s structural}}$
$+1$	-1	-2	$= -2\,[-2 + 2 \times (1 - 1)] = +4 \times \sigma_{e\text{ s structural}}$

Table 3.3.

An example in RFID (remotely powered tag)

In general and specific terms, in order to modulate the area $\sigma_{e\,s}$ of the tag, the value of R_l is reduced by what is known as "load modulation", or else we modify the value of the tuning capacitance upward or downward, using a variable capacitance diode; very often, it is not possible to modify both of these simultaneously, as the resistive part of the load $R_l = R_{\text{ant t}}$ takes a value of $(R_{\text{ant t}} - dR)$ and its reactive part $X_l = -X_{\text{ant t}}$ takes a value of $(-X_{\text{ant t}} + dX)$. Let us return to the original general equation for the current I flowing through the equivalent circuit:

$$I = \frac{1}{(R_{ant\,t} + R_l) + j(X_{ant\,t} + X_l)} V_{equi}$$

and multiply the numerator and denominator by the conjugated quantity of the denominator:

$$I = \frac{(R_{ant\,t} + R_l) - j(X_{ant\,t} + X_l)}{(R_{ant\,t} + R_l)^2 + (X_{ant\,t} + X_l)^2} V_{equi}$$

Then it becomes:

$$I'' = \frac{(R_{ant\,t} + (R_{ant\,t} - dR)) - j(X_{ant\,t} + (-X_{ant\,t} + dX))}{(R_{ant\,t} + (R_{ant\,t} - dR))^2 + (X_{ant\,t} + (-X_{ant\,t} + dX))^2} V_{equi}$$

$$I'' = \frac{(2R_{ant\,t} - dR) - jdX}{(4R_{ant\,t}^2 + (dR)^2 - (4R_{ant\,t}\,dR)) + (dX)^2} V_{equi}$$

Hypothetical cases of tag operation

Let us assume that the total load impedance is only slightly different from that of the matching condition, in other words, dR is small with respect to $R_{ant\,t}$, and/or dX is small with respect to $X_{ant\,t}$, and therefore the terms $(dR)^2$, $(dX)^2$ and $(4R_{ant\,t}\,dR)$ of the equation above are negligible with respect to the others.

Thus, we obtain:

$$I'' = \frac{(2R_{ant\,t} - dR) - jdX}{4R_{ant\,t}^2} V_{equi}$$

and therefore, finally:

$$I'' = \frac{V_{equi}}{4R_{ant\,t}} \left[\left(2 - \frac{dR}{R_{ant\,t}} \right) - j\frac{dX}{R_{ant\,t}} \right]$$

Now let us calculate the new effective RCS $\sigma''_{e\,s}$ during this phase of impedance modulation. We can do this by using the same type of calculation as before.

We know that, by definition, the new power reradiated by the tag P''_s is equal to:

$$P''_s = (R_{ant\,t} I''^2) G_{ant\,t}$$

Since we know the new complex value of I'':

$$I'' = \frac{(2R_{\text{ant t}} - dR) - jdX}{4R_{\text{ant t}}^2} V_{\text{equi}}$$

we can calculate its effective value (in other words, the value of its modulus):

$$|I''| = \frac{\sqrt{(2R_{\text{ant t}} - dR)^2 + (dX)^2}}{4R_{\text{ant t}}^2} V_{\text{equi eff}}$$

and then square it:

$$|I''|^2 = \frac{(2R_{\text{ant t}} - dR)^2 + (dX)^2}{16R_{\text{ant t}}^4} V_{\text{equi eff}}^2$$

In this case (transponder impedance modulation), the reradiated power P''_s will, therefore, be:

$$P''^2_s = R_{\text{ant t}} \frac{(2R_{\text{ant t}} - dR)^2 + (dX)^2}{16R_{\text{ant t}}^4} V_{\text{equi eff}}^2 G_{\text{ant t}}$$

We also know that the total structural power P_t received by the tag from the base station is:

$$P_t = \frac{\lambda^2}{4\pi} G_{\text{ant t}} s \quad \text{in watts}$$

and therefore

$$s = \frac{P_t \times 4\pi}{\lambda^2 G_{\text{ant t}}} \quad \text{in W m}^{-2}$$

and the reradiated power P''_s will now be:

$$P''_s = \sigma''_{e s} s$$

and therefore:

$$\sigma''_{es} = \frac{P''_s}{s}$$

Now we can transfer the value of s into the equation for $\sigma_{e\,s}$, then replace P''_s with its value, which gives us:

$$\sigma''_{es} = \lambda^2 G^2_{ant\,t} \frac{(2R_{ant\,t} - dR)^2 + (dX)^2}{4\pi \times 16 R^3_{ant\,t}} \frac{V^2_{equi\,eff}}{P_t}$$

We have seen that, when the impedances of the tag antenna and the load are matched, the power P_t is equal to the power dissipated in the load, i.e.:

$$P_t = \frac{V^2_{equi\,eff}}{4R_{ant\,t}}$$

If we include this value into the preceding equation, we obtain:

$$\sigma''_{es} = \lambda^2 G^2_{ant\,t} \frac{(2R_{ant\,t} - dR)^2 + (dX)^2}{4\pi \times 4 R^2_{ant\,t}}$$

If we expand the numerator, provided that dR is assumed to be small, in other words that dR^2 and dX^2 are negligible because they are of the second order with respect to the other terms of the equation, we find:

$$\sigma''_{es} = \lambda^2 G^2_{ant\,t} \frac{R^2_{ant\,t} - R_{ant\,t} dR}{4\pi \times 4 R^2_{ant\,t}}$$

$$\sigma_{es} = \frac{\lambda^2 G^2_{ant\,t}}{4\pi} \left(1 - \frac{dR}{R_{ant\,t}}\right) \text{ in m}^2, \text{ tag not matched}$$

$$\sigma_{es} = \sigma_{e\,s\,structural} \left(1 - \frac{dR}{R_{ant\,t}}\right) \text{ in m}^2, \text{ tag impedance not matched and not tuned.}$$

In conclusion, given the structure of the resulting equation, we can identify $(1 - dR / R_{ant\,t})$ with $(1 - \Gamma)^2 = (1 - 2\Gamma + \Gamma^2) \sim (1 - 2\Gamma)$ because $dR \ll R_{ant\,t,}$ and therefore the value of Γ^2 is much smaller than that of Γ, and so:

$$\Gamma = \frac{dR}{2R_{ant\,t}}$$

We could have expected this because

$$\Gamma = \frac{a-1}{a+1} \quad \text{where} \quad a = \frac{R_l}{R_{\text{ant t}}}$$

and if R_l is close to $R_{\text{ant t}}$ (and therefore dR is small), a is close to 1:

– since a is close to 1, the numerator is $(R_{\text{ant t}} + \text{d}R) - R_{\text{ant t}} = \text{d}R$;

– since a is close to 1, the denominator is $(R_{\text{ant t}} + \text{d}R) + R_{\text{ant t}} = 2R_{\text{ant t}}$ because dR is small, and therefore

$$\Gamma = \frac{\text{d}R}{2R_{\text{ant t}}}$$

A few notes

I may have spent a long time in setting out and presenting all the foregoing equations, only achieving useful approximations at the end of a series of proofs, but this has not been done in a spirit of excessive masochism. I have acted in this way purely because I wanted to allow readers and users to provide their own simplifications in the course of their calculations, in line with the specific circumstances of their RFID applications.

I should also point out that this equation does not depend on dX, and therefore, in this case, the variation of the RCS is essentially due to the load variation dR, and the phase does not change to any great extent.

Another note

I could have followed a direct argument based on Γ.

$$\Gamma = \frac{Z_l - Z_{\text{ant t}}}{Z_l + Z_{\text{ant t}}}$$

$$\Gamma = \frac{(R_l + \text{j}X_l) - (R_{\text{ant t}} - \text{j}X_{\text{ant t}})}{(R_l + \text{j}X_l) + (R_{\text{ant t}} + \text{j}X_{\text{ant t}})}$$

$$\Gamma = \frac{(R_l - R_{\text{ant t}}) + \text{j}(X_l + X_{\text{ant t}})}{(R_l + R_{\text{ant t}}) + \text{j}(X_l + X_{\text{ant t}})}$$

If we now multiply the numerator and denominator of this expression by the conjugate of the denominator, we obtain:

$$\Gamma = \frac{[(R_l - R_{ant\,t}) + j(X_l + X_{ant\,t})] \cdot [(R_l + R_{ant\,t}) + j(X_l + X_{ant\,t})]}{(R_l + R_{ant\,t})^2 + j(X_l + X_{ant\,t})^2}$$

If there is optimal matching, (the output impedance of the antenna and the input impedance of the integrated circuit are conjugate, i.e. $R_l = R_{ant\,t}$ and $X_l = -X_{ant\,t}$), then $\Gamma = 0$ and the maximum available power will be transferred to the load. Before going more deeply into the theory, let us return to the specifics of UHF and SHF RFID applications. The tag impedance will be matched and then slightly mismatched; in other words, the resistive part of the load $R_l = R_{ant\,t}$ will take the value of $(R_{ant\,t} - dR)$ and its reactive part $X_l = -X_{ant\,t}$ will move to a value of $(-X_{ant\,t} + dX)$.

By replacing R_l and X_l with their new values in the equation for Γ, with no simplification, we obtain:

$$\Gamma = \frac{\left[-2R_{ant\,t}dR + dR^2 + dX^2\right] + j\left[(2R_{ant\,t}dX\right]}{4R_{ant\,t}^2 - 4R_{ant\,t}dR + dR^2 + dX^2}$$

Assuming that the variations of resistance dR and reactance dX are small (a small detuning of the tag), in a first approximation we can disregard the terms dR^2, dX^2 and the product $(dRdX)$, which are all of the second order. After simplification, we have:

$$\Gamma = \frac{-dR + jdX}{2(R_{ant\,t} - dR)}$$

and since $dR \ll R_{ant\,t}$:

$$\Gamma = \frac{-dR + jdX}{2R_{ant\,t}}$$

Given that, when the tag is only very slightly mismatched,

$$\Delta\sigma_{e\,s} \approx (-2\Gamma)\sigma_{es\,structural}$$

$$\Delta\sigma_{e\,s} = \sigma_{e\,s\,structural}\frac{dR - jdX}{R_{ant\,t}}$$

The value in brackets is the tag modulation merit factor and shows the real and imaginary parts that may be included in the value of $\Delta\sigma_{e\,s}$.

Important note about UHF RFID systems using backscattering phase modulation

It should be noted that, based on the assumptions stated at the start of these explanations, if the value of dR is very small or even zero, and if only the value of dX is significantly modified (for example by modifying or modulating only the internal capacitance of the integrated circuit of the tag, or, in other words, by keeping Z_{ant} equal to Z_l in the forward link from the base station to the tag), the value of $\Delta\sigma_{e\,s}$ is a purely imaginary quantity. Essentially, this means that there is no change in the reradiated power in watts (backscattering) between the forward and return link phases of communication, and that only the phase of the signal reradiated by the tag is modified by the (reactive) impedance modulation of the integrated circuit. In this case, the base station receiver has to carry out a phase demodulation of the backscattering signal, instead of performing amplitude variation demodulation of the ASK type on the received power as before. By way of information, 99% of commercially available base stations, which for many other reasons have I, Q demodulators, always carry out simultaneous amplitude and phase modulation.

Matching factor

To conclude this example, let us now calculate the value of the matching factor (actually the mismatching factor) of the tag:

$$\theta_{matching} = 1 - |\Gamma|^2 = q$$

To do this, we start by calculating the modulus of this expression:

$$|\Gamma| = \frac{\sqrt{(dR)^2 + (dX)^2}}{2(R_{ant\,t} - dR)}$$

and then square it:

$$|\Gamma|^2 = \frac{(dR)^2 + (dX)^2}{4(R_{ant\,t} - dR)^2}$$

$$1 - |\Gamma|^2 = 1 - \frac{(dR)^2 + (dX)^2}{4(R_{ant\,t} - dR)^2} = q$$

NOTE.–

Theoretically, the input impedance R_l of the integrated circuit is not equal to 73 Ω. Consequently, there is always an impedance matching circuit (a transformer or an LC circuit) which adjusts the impedance represented by the integrated circuit (about 35 μW at 2 V with $P = U^2/R_l \rightarrow R_{lic} = 80$ kΩ) to 73 Ω .

3.3. Variations of $\Delta\sigma_{e\,s} = f(a)$

For reasons of simplicity, we often prefer to consider the variations of $\Delta\sigma_{e\,s}$ not as a function of Γ, but as a function of $a = R_l/R_{ant}$. We can do this by returning to the previous general equation, $\Delta\sigma_{e\,s} = f(\Gamma)$ and replacing Γ with its value as a function of a. Given that:

$$\sigma_{e\,s} = \frac{4}{(a+1)^2}\, \sigma_{e\,s\,structural}$$

we can now calculate $\Delta\sigma_{e\,s} = (\sigma_{e\,s} - \sigma_{e\,s\,structural})$, for the case of optimal matching:

$$\Delta\sigma_{e\,s} = \sigma_{e\,s} - \sigma_{e\,s\,structural} = \left(\frac{4}{(a+1)^2} - 1\right)\sigma_{e\,s\,structural}$$

Now we are "theoretically" ready to apply the backscattering principle to RFID technology but first of all we had better confront the cold hard facts, if only for a few moments.

3.4. After the theory, RFID at UHF and SHF realities

We have now concluded our study of the theory of $\sigma_{e\,s}$/RCS and other forms of $\Delta\sigma_{e\,s}$ and ΔRCS, which is absolutely necessary, but not sufficient. Everything we have said about the variations of σ_{es}/RCS relates to the best possible world, in which the tag can effortlessly handle very small fields E and H (when the tag is at a long distance from the base station) and very large fields E and H (when the tag is located very near the base station).

In fact, if the tag is going to be able to operate correctly at both long and short distances, we must include a new element in the integrated circuit, namely a "shunt regulator", to ensure that excessive voltages do not appear across its terminals in the presence of strong fields.

Everything discussed so far relates to the phase of operation when this regulator is inactive, which is important, since it defines the maximum operating distance of the system. As the tag approaches the base station, the shunt regulator increasingly comes into action in parallel at the input of the integrated circuit, thus mismatching the load from the radiation resistance of the antenna R_{ant}, causing a displacement of the reference point (where there is no modulation) which, therefore, approaches the vertical axis of the curve, making the value of ΔRCS smaller than before. This makes it harder to guarantee the variation ΔRCS for operation in proximity.

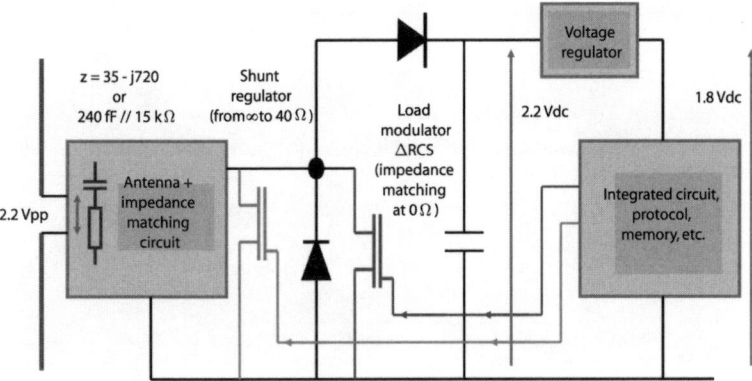

Figure 3.6. *UHF tag: block diagram (whole circuit)*

It is worth examining these matters in detail for remotely powered tags (used in most applications, therefore mainly with $\Gamma_1 = 0$), as this will help us understand the phenomena occurring in RFID at UHF and SHF. In order to do this, we need to examine the true equivalent electrical circuit of the tag, while also considering the effects of its distance from the base station.

Tag located very far from the base station

In this case, the tag does not receive (does not capture) enough energy to be remotely powered, so nothing happens, and the problem is solved.

Tag exactly on its operating threshold (maximum distance)

The tag begins to receive the exactly correct minimum amount (threshold) of power. The shunt regulator does not come into action (R_{shunt} is infinite), meaning that the tag benefits from all the possible incident energy supplied by the wave from the base station (Figure 3.7)

The "load modulation transistor" comes into action as required when data need to be sent from the tag to the base station, and switches the load impedance of the tag antenna and moves, in the case of a remotely powered tag, from the "matched" value ($a_{\text{non mod}} = R_l/R_{\text{ant t}} = 1$) (in order to obtain the maximum power) to the value $a_{\text{mod}} = (R_{\text{match}}$ in parallel with $R_{\text{mod}})/R_{\text{ant t}}$), enabling the initial backscattering area $\sigma_{\text{structural}}$ to be modified to a higher value ($\sigma_{e\,s}/\text{RCS}$) in order to reradiate more of the incident wave. The graphs in Figure 3.8 summarize the variation of ΔRCS in the case of very weak fields.

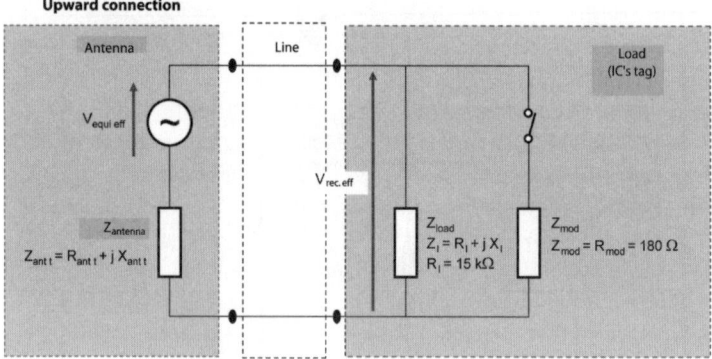

Figure 3.7. *Tag in the threshold field*

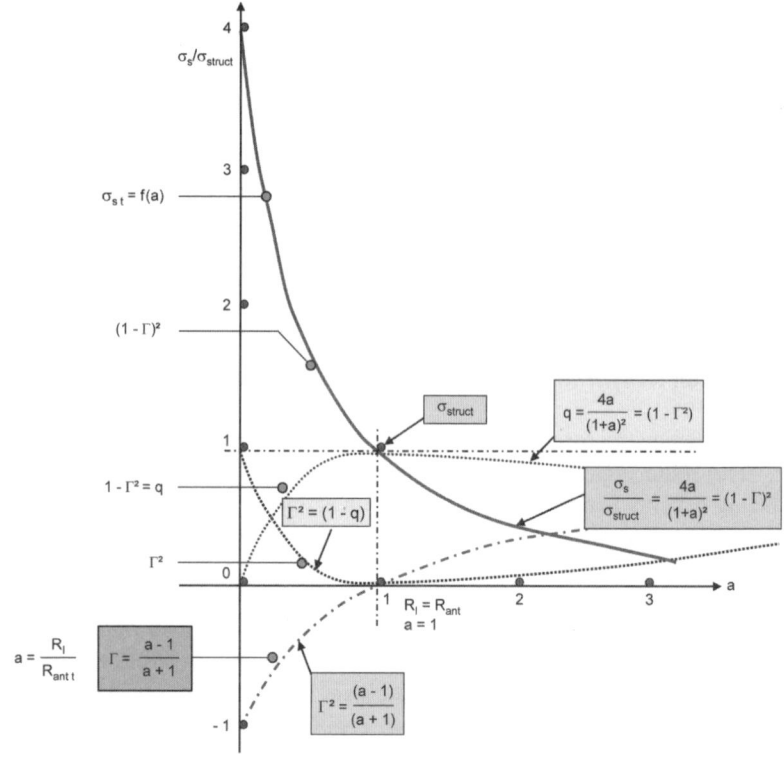

Figure 3.8. *Graph showing the variation of ΔRCS in the case of very weak fields*

The resistance R_{mod} must be chosen in such a way that it allows the base station to understand and interpret the data (in the form of power variations) transmitted by the tag, and therefore the value of ΔRCS shown in the diagram must be above a minimum level. Consequently, there is a maximum ohmic value of $R_{mod\ max}$, and therefore of a_{mod} and Γ_{mod}. An example of a calculation is given below for UHF applications according to ISO 18000-6 (UHF) and 18000-4 (2.45GHz), in other words with ΔRCS_{min} of 50 cm^2 (see the example below).

Of course, we could consider making life simpler by having R_{mod} equal to 0 Ω, thus immediately providing (see Figure 3.9) the greatest possible variation of the area ΔRCS, provided that this has no effect on the energy consumed by the tag – but that is another story.

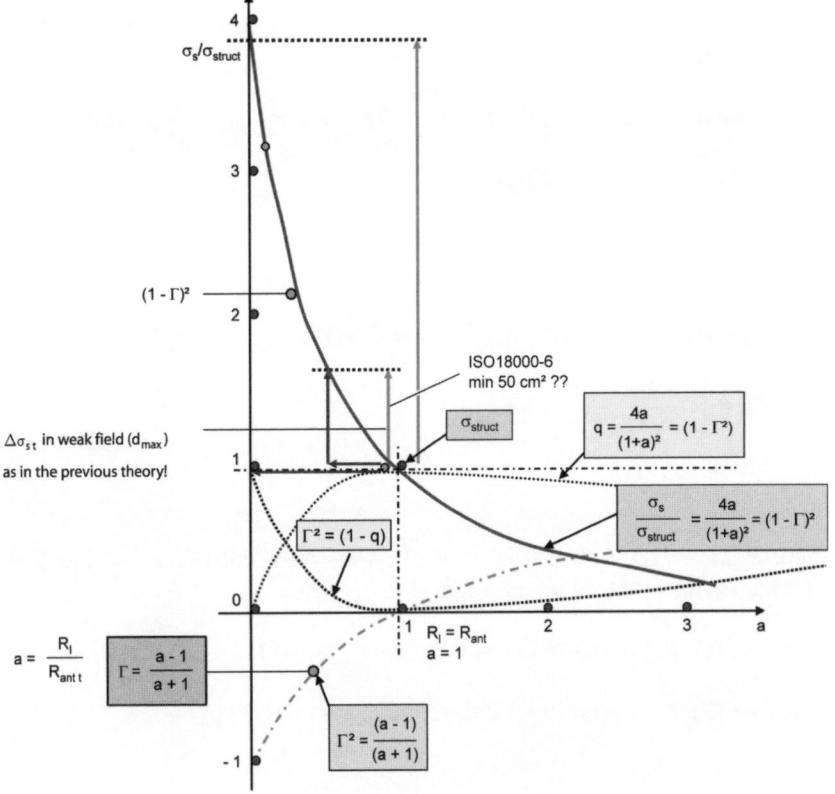

Figure 3.9. *Optimizing the variation of the load resistance for the highest value of ΔRCS*

Example: $\Delta\sigma_{es}$ and conformity with ISO 18000-6 in the threshold field

To ensure high compatibility of tags and base stations, the "Tag Parameter: 7d" of ISO 18000-6 states that "the tag ΔRCS (Varying Radar Cross Sectional area) affects system performance. A typical value is greater than 0.005 m^2 = 50 cm^2." This means that, regardless of the frequency within the 860–960 MHz band (the mandatory range for conformity with ISO 18000-6), the minimum typical value of $\Delta\sigma_{es}$ is 50 cm^2. Additionally, given that the value of $\sigma_{e\,s\,structural}$ is 0.214 λ^2 for a $\lambda/2$ dipole tag antenna with a gain of 1.64, and that

$$\Delta\sigma_{es} = \left(\frac{4}{(a+1)^2} - 1 \right) \sigma_{e\,s\,structural}$$

we can estimate the minimum value of $a = R_l/R_{ant\,t}$ to satisfy the equation $\Delta\sigma_{es} = 50$ cm^2.

a) If the frequency = 960 MHz, $\lambda = 0.3125$ m and $\sigma_{e\,s\,structural} = 209$ cm^2:

$$50 = \left(\frac{4}{(a+1)^2} - 1 \right) \times 209$$

and therefore $a = 0.797$;

b) If the frequency = 860 MHz, $\lambda = 349$ m and $\sigma_{e\,s\,structural} = 260$ cm^2:

$$50 = \left(\frac{4}{(a+1)^2} - 1 \right) \times 260$$

and therefore $a = 0.83$

where $R_{ant\,t} = 73.128$ Ω and $R_l = aR_{ant\,t}$, with these conditions we can find the maximum value of the minimum value of R_l.

In case (a): $R_{l\,min} = 58.28$ Ω and therefore $\Gamma_{min} = -0.113$.

In case (b): $R_{l\,min} = 60.7$ Ω and therefore $\Gamma_{min} = -0.093$.

NOTE.–

With these very low values of Γ_{min} (about –0.1), we are in the area where Γ can be expected to approach zero; thus:

$$\Delta\sigma_{es} = -2\Gamma\sigma_{e\,s\,structural}$$

and therefore

$$\Delta\sigma_{e\,s}= (-2) \times (-0.1) \times 250 = +50 \text{ cm}^2 \text{ Q.E.D.}$$

Tag entering its normal operating range

The tag is now correctly supplied with power, even above its strict minimum level. The shunt regulator comes into action (or comes back into action) to limit the voltage at the input of the integrated circuit (Figure 3.10).

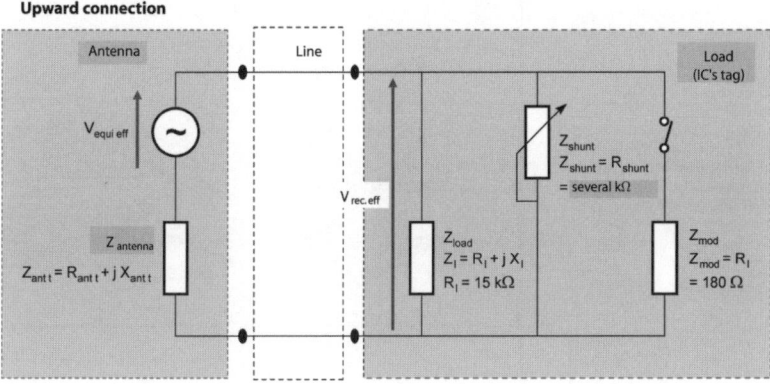

Figure 3.10. *Tag in the medium field – the shunt regulator starts to conduct normally*

For this purpose, the value of R_{shunt} decreases (by a few kilo-ohms) and is connected directly in parallel with the load resistance of the antenna, thus causing a structural mismatch of the antenna, even if there is no modulation by the modulation transistor, and a decrease of the nominal value (without modulation) of a which thus becomes $a' = (R_l$ in parallel with $R_{shunt})/R_{ant\,t}$. This is equivalent the displacement of the initial operating point of $a = 1$, $\Gamma_1 = 0$ (the point corresponding to the operating threshold) toward $a' < 1$ and Γ_1 other than zero and with a slightly negative value (Figure 3.11).

According to the data to be transmitted to the base station, the modulation transistor of the tag comes into operation and switches the above value (a') (which is no longer the matched value, so the maximum power is no longer received, although this does not upset the tag because it is closer to the base station) to the value R_{mod}, thus modifying the backscattering area (RCS) in order to reradiate more of the incident wave.

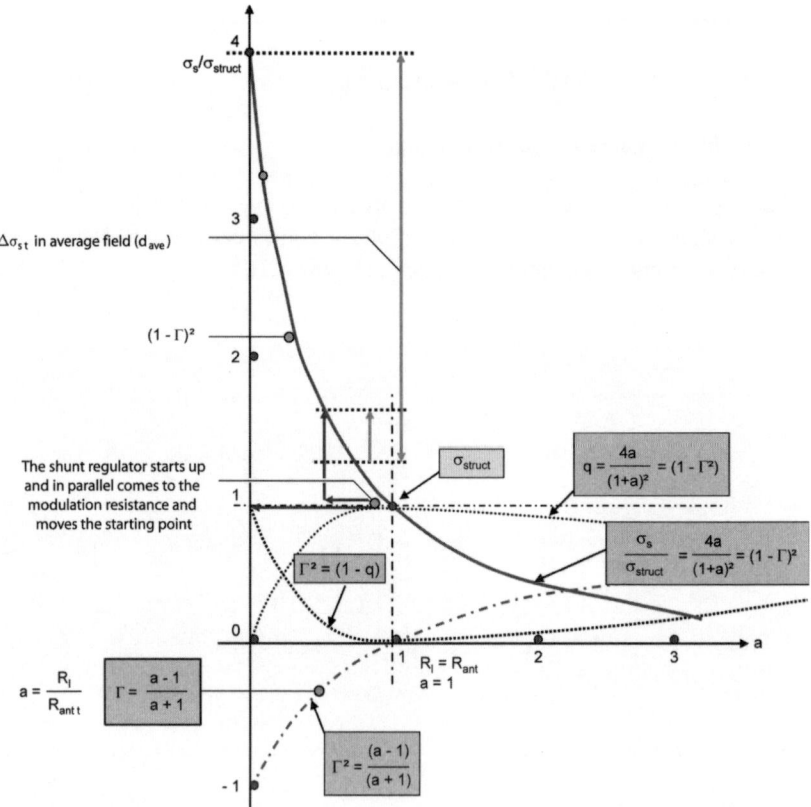

Figure 3.11. *Variations of the position of Γ_1 as a function of the field strength*

As shown in the diagram, this modulation changes the value of a, but also causes a further variation of ΔRCS, which is smaller than the previous variation. Furthermore, the value of ΔRCS decreases as the tag approaches the base station. The question which then arises is: will we fall below the minimum ΔRCS required by the standard? To answer this question, we must consider the most unfavorable limiting case, which is present when...

The tag is located extremely close to the base station

The tag is correctly powered, even well above its absolute minimum level (Figure 3.12).

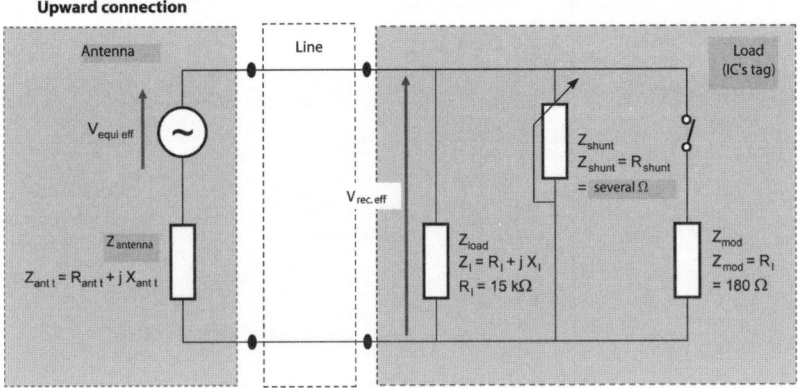

Figure 3.12. *Tag in a strong field – the shunt regulator is fully conducting*

The shunt regulator is in full operation, thus limiting the voltage at the input of the integrated circuit. For this purpose, the value of R_{shunt} becomes very low (at a few ohms to a few tens of ohms) and is connected directly in parallel with the load resistance of the antenna, thus causing a very large structural mismatch of the antenna, even if there is no modulation by the modulation transistor, and therefore a very large decrease of the nominal value (without modulation) of a which thus becomes $a'' = (R_l$ in parallel with $R_{shunt})/R_{ant\ t}$. This is equivalent to shifting the initial operating point (at the operating threshold) of $a = 1$ toward a'' which is located very close to the vertical axis in Figure 3.13, and bringing Γ_1 close to the value of -1.

According to the data to be transmitted to the base station, the modulation transistor continues to attempt to operate, switching the value a'' to the value $(R_{adapt}//R_{shunt}//R_{mod} \approx R_{shunt}//R_{mod})$ to modify the backscattering area (RCS) as much as possible in order to reradiate some of the incident waves.

As shown in the diagram, this modulation not only changes the overall value of a and Γ, but also causes a further variation of ΔRCS, which is even smaller than the previous variation. Being very close to the vertical axis, the value of ΔRCS is even smaller, but must be greater than the minimum value of ΔRCS required by the standard. This limits both the minimum range and the value of R_{shunt}, and therefore the maximum strength of the field E that the tag can accept. Also, for simple reasons of power dissipation, chip manufacturers specify a maximum input current for their integrated circuits, for example 10 or 30 mA eff (refer to the first two chapters of this book, if necessary, for more information).

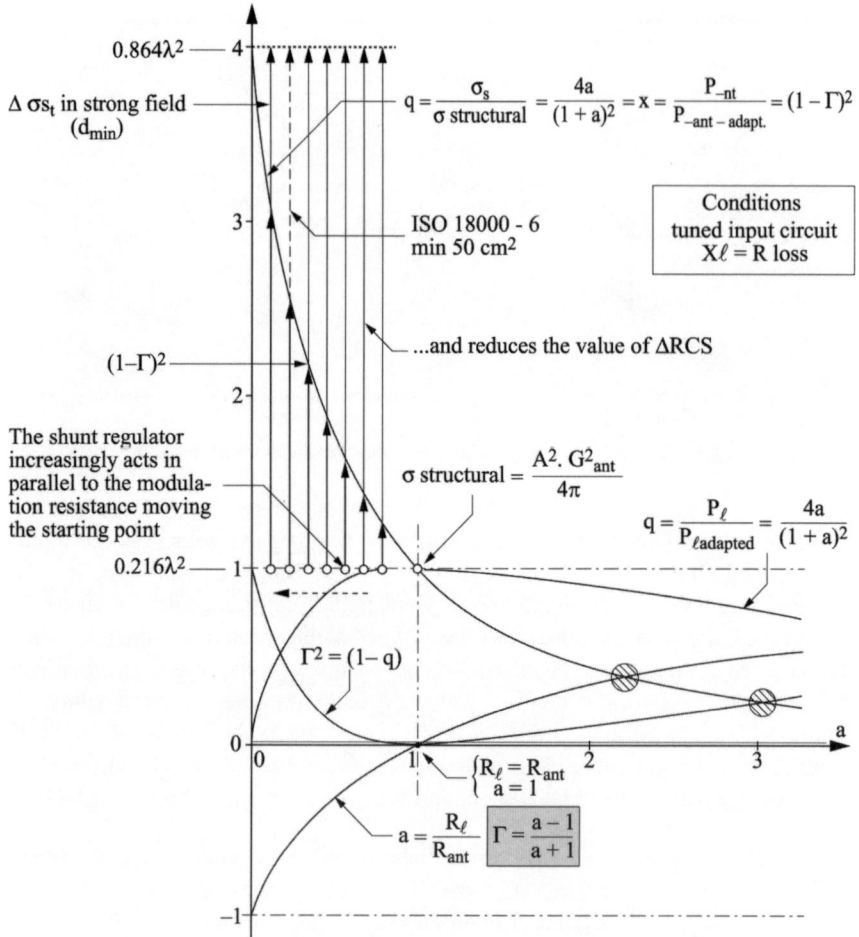

Figure 3.13. *The limiting case of the position of Γ_1 in a strong field*

Example of a real case and conformity with ISO 18000-6 in strong fields

As we have just seen, as the tag approaches the base station the shunt regulator comes increasingly into operation, and the value of Γ_1 (in the phase without modulation of the antenna load) tends toward -1.

If we wish to comply with ISO 18000-6, even in strong fields, the value of $\Delta\sigma_{e\,s}$ $_{typical}$ must be equal to 50 cm^2. Using a $\lambda/2$ dipole in which the value of $\sigma_{e\,structural}$ is approximately 200 cm^2 at 900 MHz, and short-circuiting the antenna entirely in the modulation phase, our aim is to obtain $\Gamma_2 = -1$ during the modulation. We can use the equations shown above to determine the critical value of Γ_1 corresponding to this condition:

$$\Delta\sigma_{e\,s} = \sigma_{e\,s2} - \sigma_{e\,s1}$$

$$\Delta\sigma_{e\,s} = \Delta\Gamma\,[-2 + (\Gamma_2 + \Gamma_1)]\sigma_{e\,structural}$$

$$\Delta\sigma_{\_} = [-2(\Gamma_2 - \Gamma_1) + (\Gamma_2 - \Gamma_1)(\Gamma_2 + \Gamma_1)]\,\sigma_{e\,structural}$$

Transferring the values:

$$50 = [-2(-1 - \Gamma_1) + (-1 - \Gamma_1)(-1 + \Gamma_1)]\times 200$$

we obtain a second degree equation in Γ_1, $[-\Gamma^2_1 + 2\Gamma_1 + 2.75]$, whose roots are

$$\Gamma'_1 = \frac{-2 + 3.873}{-2} = -0.89$$

$$\Gamma''_1 = \frac{-2 - 3.873}{-2} = -2.93 \text{ , a physically impossible value}$$

that is $\Gamma_1 = -0.89$ (tag unmodulated, in a strong field, with shunt fully operational).

We can now determine the value of a as follows:

$$a = \frac{1 + \Gamma}{1 - \Gamma}$$

$$a = \frac{1 - 0.89}{1 + 0.89}$$

$$a = 0.058 = R_l/R_{ant\,t}$$

which shows that, in order not to exceed the minimum value required by ISO 18000-6, the minimum value of the shunt resistance must not fall below $R_{shunt\,min} = 0.058 \times 73 = 4.24\ \Omega$ (with $R_{ant\,t}$ of 73 Ω on matching). The position of this point is shown in Figure 3.13 above.

By way of conclusion

As shown above, the value of $\Delta\sigma_{e\,s}$ decreases as we approach the base station. Precise measurements show that the measured variation of $\Delta\sigma_{e\,s}$ is substantially proportional to r^2. Additionally, the variation of the power reradiated by the tag, ΔP, is equal to $s\Delta\sigma_e\ _s$. Now, the incident power flux density s is equal to $\dfrac{P_{eirp}}{4\pi r^2}$, indicating that the ΔP reradiated by the tag is substantially constant regardless of the distance at which it operates; this property being mainly due to the presence and operation of the shunt regulator.

This concludes the major theoretical part of this section of the chapter which is primarily concerned with reflection, absorption and the backscattering transmission principles and their effects. Unfortunately, as you will certainly have realized, the concepts of RCS and $\Delta\sigma_e\ _s$ (or ΔRCS) are rather difficult to grasp, and the values involved are not very easy to measure.

3.5. Measuring ΔRCS

As I have shown above, when the tag circuit modulates the value of RCS, this is done by using a transistor operating in switch mode, in other words on an "on/off" basis, and therefore using square wave signals (with or without subcarriers) which, when antenna mismatching takes place, produce a frequency spectrum including sidebands that are located on either side of the carrier frequency and that thus represent the modulating signal.

Most of the energy of the signal reradiated by the tag and representing the transmitted data is therefore in these sidebands, and this is also the location of the power present in the return signal, since, if we keep strictly to the conceptual aspect of the RCS (measurement of the return power contained in the incident carrier only) we may find it very difficult to recover the signal. Also, we can use return modulation of the Manchester subcarrier coding (SCM) or BPSK type, as for RFID operating at 13.56 MHz (ISO 14443 and 15693), in an attempt to escape from the use of the unwieldy carrier signal for the amplification and demodulation of the very weak return signal.

3.5.1. Example of a method for measuring ΔRCS

Returning to our topic, we can obtain the value of $\Delta\sigma_e\ _s$ (often called ΔRCS) by using the measurement setup shown in Figure 3.14, in which a base station supplies/transmits a constant isotropic power $P_{bEIRP} = P_{cond}G_{ant\ bs}$. The power flux density radiated by the base station and present at the tag, at a distance r_1 from the base station, is, as expected,

$$a = \frac{P_{eirp}}{4\pi r^2}$$

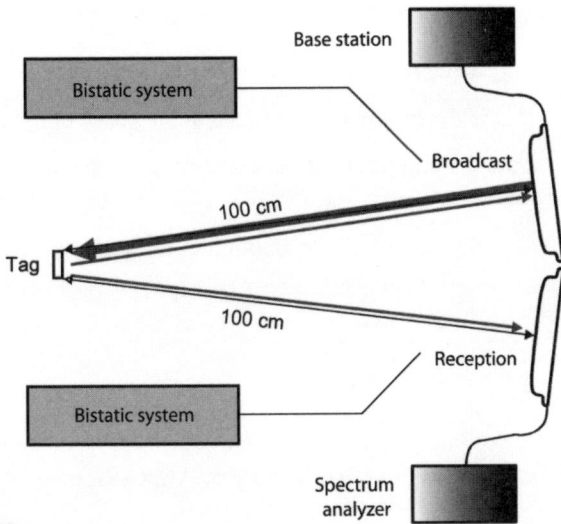

P₁   power from the base station

P_1 power from the base station
G_s broadcast antenna gain from the base station
R_1 distance between the tag and the base station broadcast antenna
R_2 distance between the tag and the measurement receiver antenna
G_R gain of the receiving antenna of the measurement receiver
$P_{s,1}$ power measured in the first lateral band of the spectrum
λ wavelength of the carrier wave

Figure 3.14. *Method of measuring ΔRCS*

Let $P_{s\ \text{tag EIRP}}$ denote the (difference in) global power EIRP reradiated by the tag when the return wave is modulated by modulation of the tag impedance using a square wave signal with a peak amplitude h. This is analyzed in a conventional way into the Fourier series:

$$f(x) = \frac{4h}{\pi} \cdot \left(\cos x - \frac{\cos 3x}{3} + \frac{\cos 5x}{5} - \ldots \right)$$

indicating that the amplitude of the first harmonic (the fundamental frequency) of the function that produces it is greater than $4/\pi$ ($= 1.27$) at the initial value h of the square wave signal. Also, assuming that this square wave signal creates an amplitude modulation (AM) of the incident UHF/SHF carrier, the two reradiated sidebands created by this modulation support the signal.

Given that:

– U_{max} is the maximum amplitude in the phase of dynamic modulation of the tag;

– U_{min} is the minimum amplitude in the phase of dynamic modulation of the tag;

– U_b is the amplitude of the carrier present when there is no dynamic modulation of the tag;

– U_c is the sum of all the amplitudes of the signals making up the square wave;

we can write:

a) $U_{max} = U_c + 2\ U_h$

b) $U_{min} = U_c - 2\ U_h$

and $U_{min} = 0$, when we perform 100% AM (ASK).

If we now simplify matters by identifying the real square wave signal U_h with its first harmonic and call the amplitude of its first harmonic (fundamental) $U_{s,1}$, Fourier analysis gives us:

$$Uh = U_{s,1} \times \frac{\pi}{4}$$

and if we transfer this value into the above equations, we obtain:

$$U_{max} = U_c + 2U_{s,1} \times \frac{\pi}{4}$$

$$U_{min} = U_c - 2U_{s,1} \times \frac{\pi}{4}$$

Since U_{min} is equal to 0 in AM of the 100% ASK type, we can combine the last two equations to give:

$$U_{max} = U_{s,1} \times \pi$$

This voltage-based description can be rephrased in terms of power (proportionally to the square of the voltage). This entails a relationship such that the power reradiated by the tag (and therefore received by any receiver) P_{max} corresponding to U_{max} is equal to:

$$P_{max} = P_{s,1} \times \pi^2$$

where $P_{s,1}$ is the power contained in the first sideband reradiated by the tag, which is about 10 times greater than that expected from a purely static (or slow) modulation of the tag impedance, provided that the return signal is only read by a detector whose analysis aperture has a narrow bandwidth (a few kHz) to ensure that only the sideband(s) due to the modulation signal are observed. Figure 3.15 provides an idea of the spectrum reradiated by the tag as a result of modulation by a conventional square wave signal.

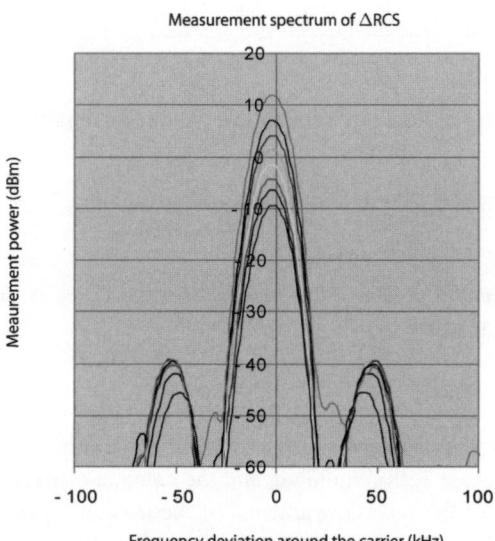

Figure 3.15. *Example of a spectrum of the reradiated power*

If we only consider the reradiated power $P_{s,1}$ contained in the first harmonic of the spectrum, this will act as an equivalent isotropic transmission source for the backscattering signal, $P_{s,1 \text{ tag EIRP}}$, part of which, P_{recept}, is recovered (in the first harmonic of the spectrum) at the receiver with a gain of $G_{\text{ant recept}}$ located at a distance r_2, according to the Friis equation:

$$P_{\text{recept}} = P_{s1 \text{ tag eirp}} \left(\frac{\lambda}{4\pi r_2} \right)^2 G_{\text{ant recept}}$$

and therefore:

$$P_{\text{s1 tag eirp}} = P_{\text{recept}} \left(\frac{4\pi r_2}{\lambda} \right)^2 \frac{1}{G_{\text{ant recept}}}$$

By definition, the variation of the RCS of the tag, ΔRCS, representing the ratio between retransmitted power and the incident power flux density will be:

$$\Delta\text{RCS} = \frac{P_{\text{s modul}} - P_{\text{s non modul}}}{s} = \frac{P_{\text{max}}}{s} = \frac{P_{\text{s1 tag eirp}} \times \pi^2}{s}$$

Combining the last of the above equations, we finally obtain:

$$\Delta\sigma_{es} = \Delta\text{RCS} = P_{\text{recept}} \times \pi^2 \frac{(4\pi)^3 r_2^2 r_1^2}{\lambda^2} \frac{1}{P_{\text{bs eirp}} G_{\text{ant recept}}}$$

P_{cond} is the conducted power of the base station and $G_{\text{ant bs}}$ is the gain of the base station transmitting antenna:

$$P_{\text{bs eirp}} = P_{\text{cond}} G_{\text{ant bs}}$$

where r_1 is the distance between the tag and the transmitting antenna of the base station, r_2 is the distance between the tag and the antenna of the measuring receiver, $G_{\text{ant recept}}$ is the gain of the receiving antenna of the measuring receiver, P_{recept} is the power received and measured in the first sideband of the spectrum and λ is the wavelength of the carrier wave.

By applying this last formula to the measured value P_{recept}, we can find the value of $\Delta\sigma_{es}$ which can be used in practice.

Note that all of these proposed measurement methods have been accepted as the basis for the measurement of ΔRCS in ISO 18047-6 entitled "Conformance Tests" for RFID at UHF.

By way of information, an example of values of $\Delta\sigma_{e\,s}$ (measured and then calculated) as a function of the power contained in the first harmonic of the switched signal is shown in Figure 3.16.

As shown above, the value of $\Delta\sigma_{e\,s}$ becomes increasingly difficult to determine as we approach the base station. "Up to what point?" the reader may well ask. Figure 3.17 shows a few "bad examples".

Note that there are two remarkable facts about this diagram:

– The horizontal axis is graduated according to the maximum distance for operation (100%) of the tag.

– The diagram shows the value of $\Delta\sigma_{e\ s\ min}$ which must be conformed with, according to ISO 18000-6.

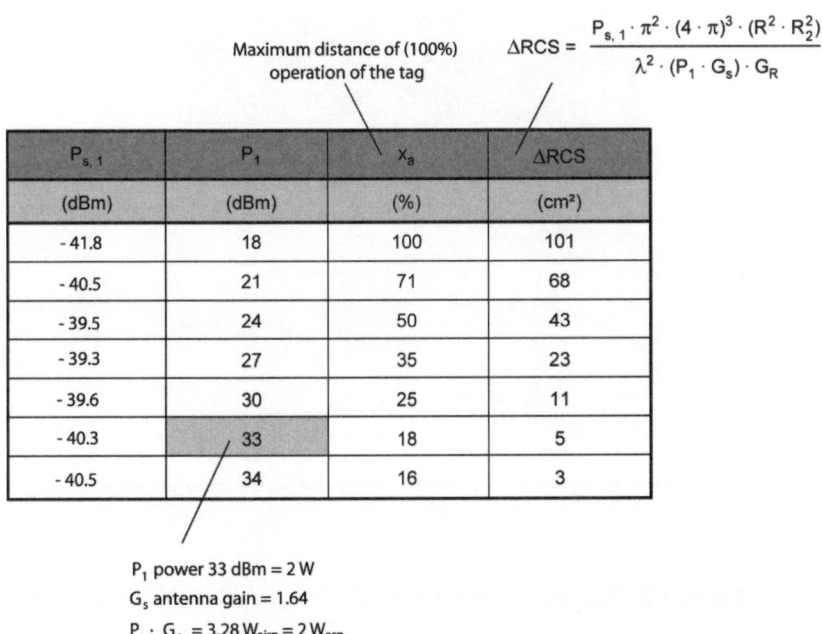

$$\Delta RCS = \frac{P_{s,1} \cdot \pi^2 \cdot (4 \cdot \pi)^3 \cdot (R^2 \cdot R_2^2)}{\lambda^2 \cdot (P_1 \cdot G_s) \cdot G_R}$$

Maximum distance of (100%) operation of the tag

$P_{s,1}$	P_1	x_a	ΔRCS
(dBm)	(dBm)	(%)	(cm²)
– 41.8	18	100	101
– 40.5	21	71	68
– 39.5	24	50	43
– 39.3	27	35	23
– 39.6	30	25	11
– 40.3	33	18	5
– 40.5	34	16	3

P_1 power 33 dBm = 2 W
G_s antenna gain = 1.64
$P_1 \cdot G_s$ = 3.28 W_{eirp} = 2 W_{erp}

Figure 3.16. *Example of measured and calculated values of $\Delta\sigma_{e\ s}$*

To finish with this subject, one should always remember that, as mentioned above, because of the necessary presence of a regulator in the tag to enable it to operate correctly in weak (far) fields and strong (very near) fields, the return modulation index will also depend on the distance, and therefore it is not as simple as it may appear to ensure a minimum value of ΔRCS.

Developers of UHF and SHF base stations have therefore familiarized themselves with all the methods of signal amplification, selection and processing, and, if this is not your main area of expertise, I suggest that you consult the many specialist books about HF signals – or, if all else fails, contact the authors.

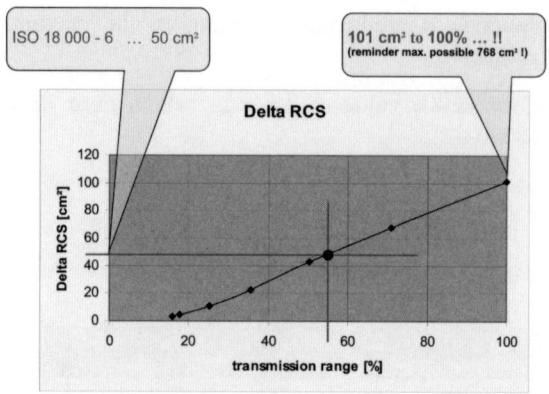

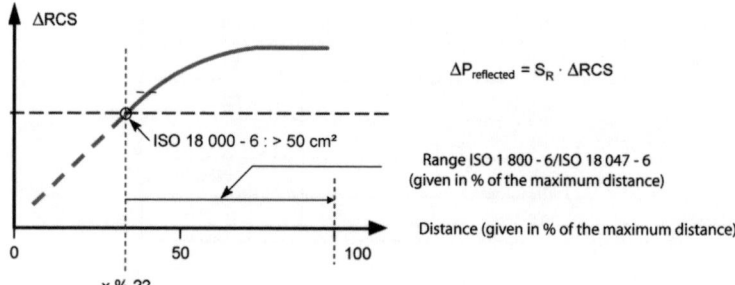

Figure 3.17. *Examples of measured values of $\Delta\sigma_{es}$ in commercial tags*

3.6. The "Radar" equation

Finally, let us now consider the radar equation.

Power reradiated by the tag and power received by the base station

Remember that, as a general rule,

$$\Delta\sigma_{es} = \Delta\Gamma\,[\Delta\Gamma + 2(\Gamma_1 - 1)]\sigma_{es\,structural} = f(\Delta\Gamma \text{ and } \Gamma_1)$$

whose value depends on both the variation of $\Delta\Gamma$ and the initial point of this variation Γ_1; therefore, when the load impedance of the tag antenna circuit is modulated, the difference in power reradiated by the tag, ΔP_{back}, between the "non-modulation" and "modulation" phases will be:

$$\Delta P_{back} = P_{back\ mod} - P_{back\ non\ mod}$$

$$\Delta P_{back} = [\Delta\Gamma \{\Delta\Gamma + 2(\Gamma 1 - 1)\}]\, P_{s\ structural}$$

and the difference in power received at the base station (allowing for the gain of its antenna) will be:

$$\Delta P_{back\ received} = [\Delta\Gamma\, [\Delta\Gamma + 2\,(\Gamma 1 - 1)]]\, P_{s\ structural} \bullet attenuation \bullet G'_{bs}$$

$$\Delta P_{back\ received} = [\Delta\Gamma\, [\Delta\Gamma + 2\,(\Gamma_1 - 1)]]\, P_{bs\ eirp}\, \frac{\lambda^2}{(4\pi r)^2} G_{ant\ t}^2 \left(\frac{\lambda}{4\pi r}\right)^2 G'_{bs}$$

Given that $P_{bs\ EIRP} = P_{bs\ cond}\, G'_{bs}$:

$$\Delta P_{back\ received} = P_{bs\ cond}\, G_{bs}^2 G_{ant\ t}^2 \left(\frac{\lambda}{4\pi r}\right)^4 \Delta\Gamma\, [\Delta\Gamma + 2\,(\Gamma_1 - 1)]]$$

This general equation enables us to design the receiving stages of the base station to make them capable of receiving and detecting the small fraction of power ΔP_{back} which is reradiated by the tag via the variation of the area ΔRCS (see the example in the preceding chapter).

3.7. Appendix: summary of the principal formulas

Antenna gain

For an isotropic dipole: gain = 1.5 or in dB = 10 log(1.5) = 1.76 dB.

For an isotropic $\lambda/2$ dipole: gain = 1.64 or in dB = 10 log(1.64) = 2.14 dB.

Power

$$P_{EIRP\ bs} = P_{cond\ bs} G_{ant\ bs}$$

$$P_{EIRP} = 1.64\, P_{ERP}$$

Power flux density produced by the base station

$$s = |S| = \frac{P_{out} G_{antbs}}{4\pi r^2} = \frac{dP}{d\sigma} \text{ in W m}^{-2}$$

$$s = \frac{P_{bs} G_{ant\ bs}}{4\pi r^2} = \frac{P_{EIRP}}{4\pi r^2} \text{ in W m}^{-2}$$

Effective area of the tag

$$\sigma_{e\,t} = \frac{\lambda^2}{4\pi}\ G_{\text{ant t}} \text{ in m}^2$$

Power received by the tag

$$P_t = \sigma_{e\,t}\,s \text{ in W}$$

The Friis equation

$$P_t = P_{\text{bs}}\ G_{\text{bs}}\ \left(\frac{\lambda}{4\pi r}\right)^2\ G_{\text{ant t}} \text{ in W}$$

$$P_t = P_{\text{eirp bs}}\ \left(\frac{\lambda}{4\pi r}\right)^2\ G_{\text{ant t}} \text{ in W}$$

Attenuation coefficient (in air)

$$\frac{1}{\left(\dfrac{v}{4\pi r}\right)^2} = \text{att} = \text{attenuation coefficient}$$

attenuation (in dB) = $-147.56 + 20 \log f + 20 \log r$ where f is in Hz and d is in m.

Power reflected by the tag

$$P_s = \sigma_{e\,s}\,s$$

Effective area or radar cross section

$$\sigma'_{e\,s} = \frac{\lambda^2 G^2_{\text{ant t}}}{4\pi} \text{ in m}^2 \text{ with tag tuned}$$

$$\sigma'_{e\,s} = \frac{\sigma_{e\,t}}{2} G_{\text{ant t}} \text{ in m}^2 \text{ with tag tuned}$$

Power reflected/scattered/reradiated by the transponder

$$P_s = \frac{P_{\text{bs}} G_{\text{bs}}}{2} \left(\frac{\lambda}{4\pi r}\right)^2 G^2_{\text{ant t}} \text{ in W}$$

Power flux density reradiated by the tag

$$s_{back} = \frac{P_{bs} G_{bs}}{2} \frac{\lambda^2}{(4\pi)^3 r^4} G_{ant\,t}^2 \quad \text{in W}$$

Return power received by the base station (three ways of writing the same equation)

$$P_{back} = P_{eirp\,bs}\, G_{bs}\, G_{ant\,t}^2 \left(\frac{\lambda}{4\pi r}\right)^4$$

$$P_{back} = P_{eirp\,bs}\, G_{bs}\, \sigma_{e\,s} \frac{\lambda^2}{64\pi^3 r^4}$$

$$P_{back} = P_{eirp\,bs} \left(\frac{\lambda}{4\pi r}\right)^2 G_{ant\,t}\, G_{ant\,t} \left(\frac{\lambda}{4\pi r}\right)^2 G_{bs}$$

Merit factor of the tag

$$\frac{\Delta\sigma_{es}}{\sigma_{es}} = \Delta\Gamma\,[\Delta\Gamma + 2(\Gamma_1 - 1)] \quad \text{where } \Delta\Gamma = (\Gamma_2 - \Gamma_1)$$

RFID Markets

Which business problems can RFID technology resolve? This is this key question that this chapter will try to answer

4.1. Introduction

The Internet of things (IoT) is emerging, and it will create a smarter world with objects connected to their environments and also to all of us.

There will come a day when a massive network of small intelligent devices will be aware of our needs and work for our actual benefit.

A number of technologies will enable IoT to be deployed widely, and the item-level RFID is a key technology that has already conquered a few applications.

To date, RFID has been used mainly for business-to-business (B2B) applications, but business-to-consumer (B2C) applications have started emerging with near-field communicating (NFC) already turning hundreds of millions of mobile phones into RFID readers.

The tipping point for RFID wide deployment is that consumers are using the technology for their own benefits.

4.2. Market inflection point: users

Although history never repeats itself the same way, it is possible to highlight some analogies between mobile phone, Internet and RFID.

In the early 1970s, the first e-mail was sent and the first handheld mobile phones appeared.

From there, it took 25 years to build a reliable infrastructure and another 5 years to build user-friendly tools. Also, when users start experiencing the importance of the technology to make their life simpler, this is the inflection point for fast technology deployment.

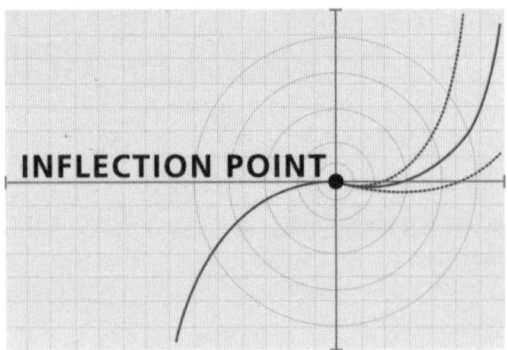

Figure 4.1.

RFID is not yet widely used, but it is heading toward this ultimate point.

4.3. RFID: what for?

An RFID user can enjoy the benefit of driving without stopping at the RFID toll collection and paying "on the fly" with RF when very long queues line up at the toll collection booths with cards and cash (see Figures 4.2 and 4.3).

RFID enables instantaneous and automatic transactions at a distance.

Figure 4.2. *Jammed toll collection*

Figure 4.3. *Fluid RFID driveway*

RFID can have more applications. It can also process multiple items at a time with several hundred per second, without line of sight (RF goes through many materials), with no battery and at extremely low cost. All these features enable an item-level RFID (see Figures 4.4 and 4.5).

Figure 4.4. *RFID cabin with up to hundreds of tags in a trolley*

Figure 4.5. *RFID manual reader with up to hundreds of tags in a case*

4.4. Open- and closed-loop applications

4.4.1. *Closed-loop applications*

In the closed-loop applications, the item cycles again and again (see Figure 4.6). Examples of this application are:

1) books in the library space;

2) garments and linen in the textile rental business;

3) gas cylinders, etc.

A service company buys items and services them. The user utilizes the item, and the owner buys and services the items when cycling them back (washing textile, filling up gas cylinders, etc.).

4.4.2. *Open-loop applications*

In the open-loop applications, the items go in a straight line, as shown in Figure 4.7.

The supply chain is a good example of an open-loop application. It goes from production up to point of sales with billions of items manufactured mainly in Asia and transported to the stores all over the world for consumers to buy. RFID is used along the supply chain enabling a reliable logistics process.

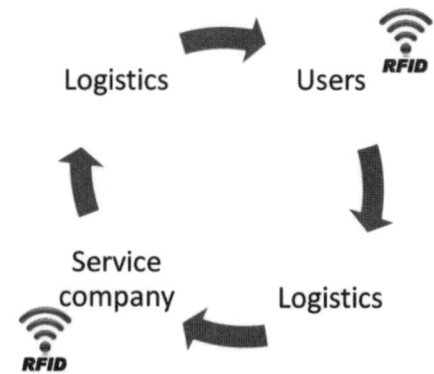

Figure 4.6. *Closed-loop application*

Figure 4.7. *Open-loop application*

4.5. RFID return on investment

4.5.1. *Introduction*

Return on investment (ROI) is at the heart of the RFID buying decision process. RFID technology must generate an ROI to trigger the associated investment.

In general, the RFID ROI is built on the cost reduction and sales increase, both generating a global ROI.

Handling millions of items moving across the world is a difficult exercise, for example, our personal laundry at home with a few hundreds of items only … but it is still a headache to pair socks (see Figure 4.8).

Figure 4.8. *Laundry at home, a headache*

Although there are many barcode systems already in place to manage the flow of items, the gap between "which items are supposed to be there" and "which items are physically there" is significant.

RFID answers the question "which items are physically there" fully automatically and instantaneously.

Fully automatically and instantaneously: professionals use RFID to automate production, distribution center (DC) and stores. Compared to barcodes, it is of course more improved in speed and accuracy, but it can be more than that: it enables operations that were impossible before, such as:

1) Control of all individual items inside a case in logistics: with barcode, it is limited to sampling control.

2) Automatic and real-time store inventory: with barcodes, it is limited to the manual inventory, very time-consuming and performed only occasionally.

Fully automatically and instantaneously: users can very simply make it intuitive, enabling self-service kiosks such as:

1) In the library space, users borrow and return books on their own, simply by putting them on a table antenna and by returning them into a chute.

2) In the laundry space, users collect their clean garments from an automatic dispenser and return the soil garments into a chute.

3) In the luggage space, travelers drop their luggage on an automatic bag drop and the luggage automatically checks in.

4.5.2. *Cost reduction*

There are many sources of cost reduction with RFID:

1) Material cost reduction, for instance:

i) Gas cylinders with the exact amount of gas inside the bottle. To fill the required quantity of gas sold, we need to know each cylinder's dead weight. With RFID, each bottle automatically declares this enabling an accurate filling. Without RFID, individual dead weights are not taken into account because of the task being too labor intensive and containing too many human errors, and gas cylinders are overfilled to ensure the minimum quantity of gas as per normal.

ii) Laundry washing machines operating in pull modes with the upstream information of the incoming items, generating higher efficiency. For instance, instead of washing a batch of "blue" items with only 60% of the machine capacity, it is preferable to wait some limited time if the information "blue items have just arrived" is available.

2) Labor cost reduction with item automatic sorting in production and logistics, for instance:

i) Garment sorters: garments have to be returned to the associated wearers. It means that a particular garment needs to be sorted out and shipped out to the right location and person. Automatic reading of the garment RFID tag enables labor-free, fast and accurate garment handling and sorting systems.

ii) Book sorters: books are returned in bulk by the users, and the librarians have to handle, sort and reposition the books on the shelves. Automatic reading of the book RFID label enables labor-free, fast and accurate book handling and sorting systems.

3) Human error cost elimination in logistics, for instance:

i) In fashion retail, DCs receive items in bulk from the manufacturers and ship the requested quantities of given items to stores. Automatic reading of the item RFID labels enables automatic check-in (incoming items from suppliers) and automatic check-out (outgoing items to the store), eliminating significant human errors.

4) Item loss identification, for instance:

i) In the textile rental industry, losses can be as high as 50% on a yearly basis with no clear responsibility between the users and the textile rental company.

Automatic reading of the item RFID tags in the laundry enables automatic check-in (incoming items from users) and automatic check-out (outgoing items to the user), and therefore the misbalance can be attributed to the user without any doubt.

4.5.3. *Sales increase*

1) In the closed-loop applications, extra services are proposed to the users, with, for instance, RFID-based kiosks. It is very easy to use because RFID reads "on its own", enabling 24/7 operation and eliminating long queues, consequently saving what we lack the most in life, i.e. time.

2) In fashion retail open-loop applications, item visibility and accurate store inventory increase sales mainly by eliminating out-of-stock situations. This starts with the RFID check-in of incoming goods when cases full of items are delivered to the store. It continues with the RFID inventory of items on the shelves and automatic overnight reordering of missing items to be ready to sell the next day. Typically, accurate store visibility generates proven sales increase between 7 and 12%.

4.6. Many RFID technologies

There are several RFID frequencies:

1) Low frequency (LF): this appeared in the mid-1980s and deployed with animals and car keys.

2) High frequency (HF): this appeared in the mid-1990s and deployed mainly in the library space.

3) Ultrahigh frequency (UHF): this appeared in the early 2000s and has been deployed in the fashion retail space; it is currently the dominant frequency.

On top of the frequency, RFID can be *passive, battery-assisted passive* (BAP) or *active*:

1) Passive means no battery at all, and the RFID chip is powered up by the RF coming from the reader.

2) BAP means a very small battery is required, just to power up the RFID chip (a few μW), not as for RF transmission (a few mW, 1,000 times more). Whether Passive or BAP, RFID chips are similar in technology; therefore, both are inexpensive and adequate for large deployments.

3) Active means a significant battery for RF transmission. It provides, of course, very large range but the battery limits the scope of applications significantly for item RFID.

RF sensing: RFID can be limited to the transmission of an ID, which is the case for most applications today. RFID can also measure physical parameters (temperature, pressure, heart beat, etc); in that case, RFID turns into RF sensing.

4.7. Examples

There are numerous RFID applications.

Let us briefly present the key applications:

RFID CAR KEY
- First major RFID deployment in the early 90s
- High anti-theft feature: the car engine can only ignite if a successful RFID transaction occurs between the RFID car key and the RFID reader at neiman level
- It is based on Low Frequency (LF) technology

CATTLE
- Each cattle head must carry a health record
- High food safety with highly secured animal ID
- It is based on Low Frequency (LF) technology

PET
- Pets must be identified, either tattooing or RFID
- It is based on Low Frequency (LF) technology

LIBRARY
- RFID enables self-service kiosk
- No more queuing up time for the user
- More time for the librarians to advice the users
- It is based on High Frequency (HF) technology

SELF SERVICE: LIBRARY RETURN CHUTE

– 24/7
– No queuing up time

SELF SERVICE: LIBRARY BORROWING STATION

– No queuing up time

LIBRARY PEDESTAL

– Antitheft

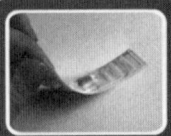

TEXTILE SERVICE

– RFID enables self-service kiosk and laundry automation
  · No more queuing up time for the user
  · Labour cost reduction
– It is based on Low, High and Ultra-High (LF, HF, UHF) technology

AUTOMATIC CHECK IN and OUT

– Roll with 100s of textile items read in an RFID cabin
– Full visibility of items delivered to and returned from customers
– Item Loss (up to 50%/y) reduces significantly

LAUNDRY AUTOMATIC SORTING SYSTEM

– Higher efficiency
– less labor cost, less errors

SELF SERVICE : RETURN CHUTE

– 24/7
– No queuing up time

SELF SERVICE : DISPENSING

– 24/7
– No queuing up time

BLOOD BAG

- Blood is a sensitive product
- RFID enables full traceability and security from the donor through the production process down to the receiver
- It is based on High Frequency (HF) technology

VIALS

- Vials can contain sensitive products
- Full traceability and security from the production down to patient
- It is based on High Frequency (HF) technology

Vial tag

©TAGSYS

BIN

- RFID bins enable "pay per weight"
- users are invoiced to the proration of what they dispose of

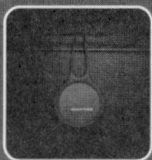

LUGGAGE PERMANENT BAG TAG

- RFID enables self-service kiosks
- No more queuing up time for the traveler
- Less labor for the Airlines
- It is based on Ultra High Frequency (UHF)

SELF SERVICE: LUGGAGE BAG DROP

LUGGAGE PAPER BAG TAG

- RFID enables efficient luggage sorting systems
- It is based on Ultra High Frequency (UHF) technology

APPAREL FASHION

- RFID enables VISIBILITY from manufacturing, through the distribution center, and to the retail floor
- It is based on the Ultra-High Frequency (UHF) technology

STORE

- Manual inventory

DISTRIBUTION CENTER

- Check in from manufacturers
- Check out to stores

4.8. Next RFID: product-embedded and seamless infrastructure

4.8.1. *Introduction*

The fashion apparel retailers are driving the large adoption of UHF RFID. It started in 2005 with Mark & Spencer in the UK with the first significant volumes (100 Mu/y), and it is deploying further right now with the US retailers, hitting larger volumes (2,000 Mu in 2013).

4.8.2. *RFID: "Slap and Ship"*

As of today, RFID for retailers is named "Slap and Slip" (Figure 4.9):

1) The RFID label, mainly hanging label, is applied to the fashion apparel product at the end of production or in the DC, and the RFID label is then discarded after sales.

2) RFID is used:

i) in logistics (DC check-in and -out), enabling accurate logistics;

ii) in the store, enabling accurate inventory.

There is no use of RFID beyond the point of sales, which means no customer interaction.

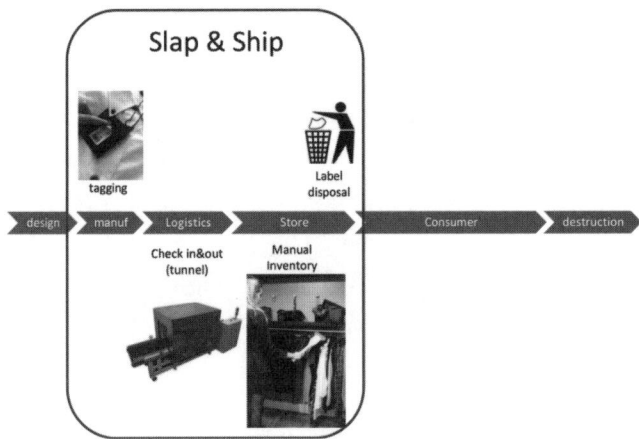

Figure 4.9. *Slap and Ship*

4.8.3. *Next RFID: from cradle to grave*

Next RFID will be used throughout the whole product life, from cradle to grave (see Figure 4.10).

It will require:

1) product-embedded RFID with the optional sensing capabilities; no more hanging labels disposed of after sales;

2) seamless and ubiquitous reading infrastructure enabling automatic store inventory; no more manual inventory with the associated human errors;

3) RFID-enabled mobile phones for the consumer to interact with the product.

4.8.4. *Embedded RFID*

4.8.4.1. *Key constraints*

Embedded RFID is a much more complex operation than just applying a hanging label onto a product.

The process must start right at the beginning of the product design to incorporate RF constraints and finally get a high-performing RFID.

Integrating an RF antenna into a product requires specific design rules as the product itself becomes part of the RF antenna.

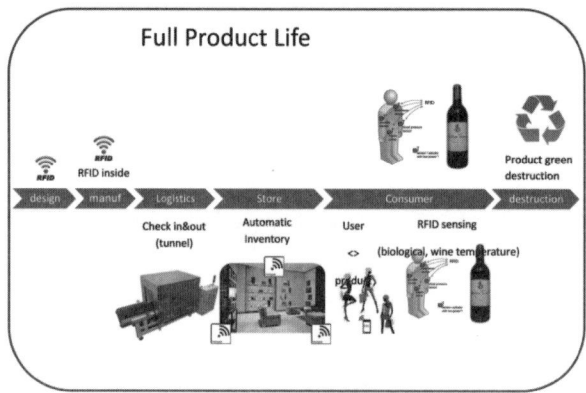

Figure 4.10. *Full lifecycle*

Key constraints are:

1) The size:

i) At UHF, the best antenna size is 15 cm, and it generates a 15 m range. Very often, it is not possible to embed such a long antenna into an item.

ii) With many zigzags, the antenna length can be reduced and the associated range will drop accordingly (see Figure 4.11).

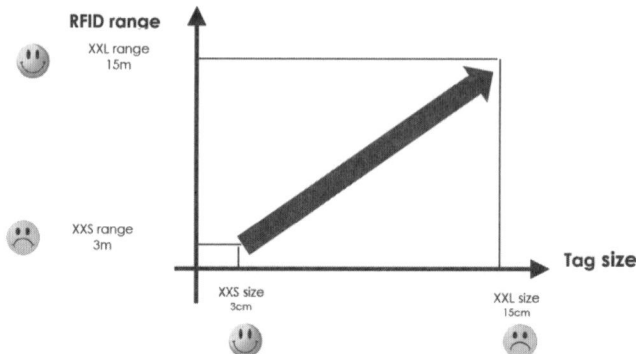

Figure 4.11. *RFID range versus tag size*

2) Item content: it is metal of course, but also usual materials such as fabrics, leather and plastic. RF is sensitive to the medium that it is applied to, and a specific design is required.

3) Ruggedization: expendable RFID labels are usually based on flip-chip technology, which is a fast and low-cost method to attach a chip onto an antenna using glue. For embedded RFID, it usually does not provide the required ruggedization. In that case, wire-bonding technology based on soldering must be used.

4.8.4.2. *A smart embedded RFID solution*

At this point, let us mention the smart concept named Stiletto, which was developed by TAGSYS, to dramatically simplify the embedded RFID.

1) It is based on the physical separation between the tag, which is very small in size, and its RF antenna, up to 15 cm.

2) The RF antenna can be carried by the item itself if this contains an adequate piece of metal; alternatively, a separate add-on RF antenna can be simply integrated into the item.

3) It enables outstanding features for embedded RFID:

i) a very small-size tag (see Figure 4.12);

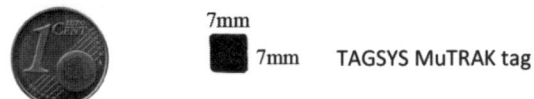

Figure 4.12. *Very small tag*

ii) at the same time, rugged and flexible; it is rugged because the MuTRAK is based on a very strong microelectronic packaging and flexible because the RF antenna can be fully flexible and follow the item deformation as much as required (see Figure 4.13).

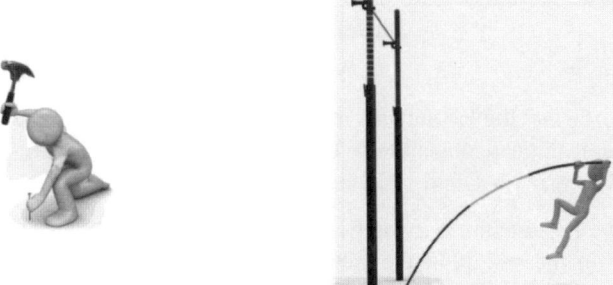

Figure 4.13. *Ruggedized and flexible tag*

4.8.4.3. *Some practical examples*

1) *Textile tag*: the RF antenna is made from a metalized thread woven into fabrics; it zigzags a little bit to reduce to overall 15 cm length. The tag, MuTRAK, is positioned in the middle of the RF antenna. The textile tag is very rugged, it survives laundry cycles and it is also fully flexible and therefore totally transparent to the wearer. It provides a range of 5 m.

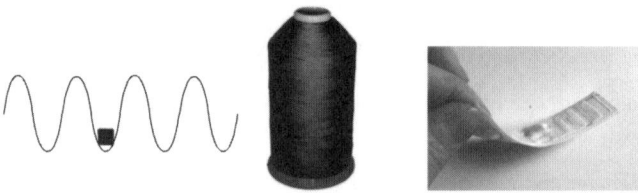

Figure 4.14. *Textile tag*

2) *High-heel shoe tag*: the RF antenna is made from the metal stiffener inside the shoe. The tag, MuTRAK, is integrated into the shoe sole in the middle of the metal stiffener. It provides a range of 5 m. This is an example where the RF antenna is found in the item itself (see Figure 4.14).

Figure 4.15. *Shoe tag*

3) *Bottle of wine*: the RF antenna is made from the metal foil on top of the bottle and the tag is positioned tangent to it. It provides a range of 3 m. This is an example where the RF antenna is found in the item itself (see Figure 4.15).

4) *Bra*: the RF antenna is made from the metallic whalebone of the bra. The tag is positioned in the middle of the metallic whalebone. It provides a range of 5 m. This is an example where the RF antenna is found in the item itself (see Figure 4.16).

Figure 4.16. *Bottle tag*

Figure 4.17. *Bra tag*

4.8.4.4. *RF Sensing*

The RF antenna is of course the heart of the RF communication.

The RF antenna can have many applications. It can also tell us about the physical parameters of the item that it is applied to. Indeed, the RF antenna is sensitive to the medium it is applied to, and the medium can itself be sensitive to the physical parameters, such as temperature, pressure and humidity.

Therefore, RF sensing enables access to the physical parameters with no extra tag component and therefore no extra cost. Examples of this include human body temperature and heartbeat, and maximum storage temperature of a wine bottle (see Figures 4.17 and 4.18).

4.8.5. *Seamless and ubiquitous infrastructure*

Embedded RFID turns items into "smart items" for their whole lifetime.

Similarly, reading devices, whether mobile or fixed, will improve to become fully non-intrusive.

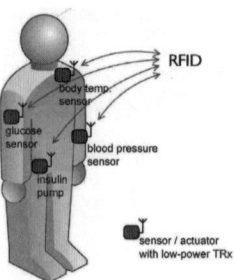

Figure 4.18. *Human heart beat and temperature*

Figure 4.19. *Storage temperature of wine bottle*

For user interaction, the mobile phone is of course the best platform (see Figure 4.19). As of today, NFC standard integrates the HF technology. UHF is likely to be integrated into NFC standards in the mid/long term.

Figure 4.20. *UHF NFC phone*

As of today, RFID-fixed antennas are pretty bulky and 100 of them are required to cover a whole store area and perform store automatic inventory. For the fixed reading infrastructure to become seamless and ubiquitous, RF technology must leapfrog on two fronts:

1) RF antenna miniaturization;

2) RF range increase to reduce the number of antennas for automatic store inventory from 100 down to 10.

Figure 4.21. *Seamless and ubiquitous infrastructure*

4.8.6. *Software for business decisions*

4.8.6.1. *Introduction*

There will come a day when a massive network of small intelligent devices will be aware of our needs and work for our actual benefit. Of course, we have not reached it, but the trend is in the right direction.

The physical world with RFID devices will generate a lot of RFID events, themselves creating the context.

From this context awareness, the logical world will provide answers to the following questions: how to react to this context? How to make the right business decisions?

4.8.6.2. *ePCGlobal middleware*

In the mid-2000s, ePCGlobal standardized the RFID middleware, inspired by the functioning of the Internet, with the following:

1) an edge server managing the RFID readers, collecting, filtering and reporting the data;

2) an electronic product code information system (EPCIS) repository to store these data structuring the what, where, when, why (i.e. "what" can be ePC number, manufacturing data (batch, expiration date, etc.), transactional data (shipment, invoice, etc.), and "why" can be: business process, such as receiving and shipping);

3) an object name server (ONS) operating as the domain name server (DNS) for the Internet and capable of finding the many EPCIS repositories (manufacturer, transporter and retailer) where a given ePC appears.

Very few ePCGlobal middlewares have emerged due to the overall system inadequacy.

4.8.6.3. *Next RFID middleware*

One complexity is coping with the huge quantities of events generated by the RFID. It starts with edge processing to keep only the pertinent data. Then, the data must be kept close to the decision point.

Finally, predictive analytics, needed to take the right decision, must combine real-time and historical data (see Figure 4.21).

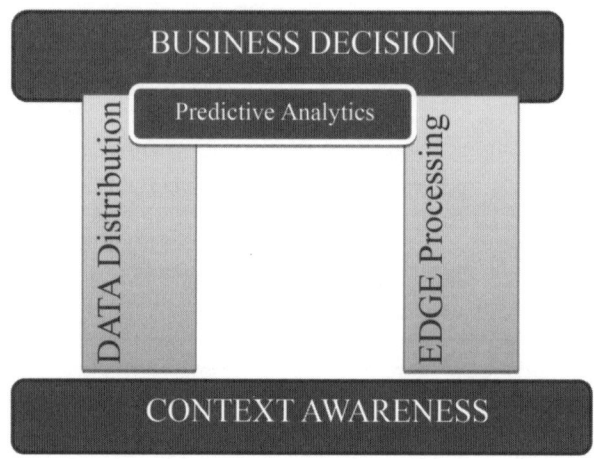

Figure 4.22. *Next RFID middleware*

Index